Surviving Collapse

Surviving Collapse

Building Community toward Radical Sustainability

CHRISTINA ERGAS

OXFORD
UNIVERSITY PRESS

Oxford University Press is a department of the University of Oxford. It furthers the University's objective of excellence in research, scholarship, and education by publishing worldwide. Oxford is a registered trade mark of Oxford University Press in the UK and certain other countries.

Published in the United States of America by Oxford University Press
198 Madison Avenue, New York, NY 10016, United States of America.

Library of Congress Control Number: 2021933680
ISBN 978-0-19-754410-5 (pbk.)
ISBN 978-0-19-754409-9 (hbk.)

DOI: 10.1093/oso/9780197544099.001.0001

1 3 5 7 9 8 6 4 2

Paperback printed by Marquis, Canada
Hardback printed by Bridgeport National Bindery, Inc., United States of America

Contents

Preface

Imagine it is the year 2050. Everything has improved since the publication of this book. The global community does better than meeting the Paris Agreement targets, and limits global warming to below 1.5°C. All nations transition away from fossil fuels and meet carbon neutral and negative goals. A climate migrant program successfully relocates people to permanent homes abroad. Poverty has been eradicated globally, and everyone is able to comfortably meet their needs. All people can expect quality care and a voice in matters that affect their lives, regardless of ability, age, gender, race, religion, or sexuality. Global indicators of health and happiness are high. Conflict, violence, and war are on the decline everywhere. You, yourself, regardless of age, are happy, healthy, and surrounded by a loving and supportive community. How did we get here? Be bold and creative. Imagine the impossible. What does it take for the world to be in this great a state? Be specific. What does the global economy look like? What are local communities doing? How are their economies shaped? Are they beholden to global markets? How has agriculture changed? How have ecosystems and Indigenous ways of knowing been protected? How is your community organized? How do you make decisions? Is everyone involved? Is everyone's input valued equally?

It may be challenging to consider a world that is completely different than the world you know, but, based on the ecological and social crises we face, a completely different world is necessary for our survival as a species. This book offers lessons from two very different communities that attempt to do the impossible work of living in opposition to the homogenous globalized arrangements of conspicuous consumption, industrial agriculture, and single-family households, and they are thriving. They maintain small environmental footprints while their quality of life remains high. Understanding how they do this is necessary if we are to comfortably survive the challenges ahead. And these examples should help to fill your well of creativity to help you envision, and begin living, an alternative future in the face of pressing catastrophe.

As this book goes to press, the year is 2020. This year began with unprecedented bushfires in Australia, whereby the whole country/continent was on fire. Over ten million hectares burned, claiming thirty-three human lives, burning over 3,000 homes, and causing the deaths and displacement of three billion animals (Richards, Brew, and Smith 2020; World Wildlife Fund 2020). A pandemic, the novel Covid-19 virus, has sickened over thirty-five million and killed over one million people, and climbing, globally. It has reorganized peoples' lives, forcing them into isolation, learning and working remotely, and it has caused mass unemployment and mass evictions as economies around the globe are in decline (Dong, Du, and Gardner 2020). Fires raging in the Amazon rainforest—called the lungs of the earth because it produces 20 percent of the world's oxygen and absorbs about 5 percent of carbon dioxide emissions—are the worst in a decade, outpacing the devastation left by the 2019 fires, which had been previously called the worst fire season since 2010. Smoke from the fires has impacted air quality, hospitalizing thousands of people with respiratory illnesses and worsening the symptoms of those afflicted with Covid-19 (Woodward 2019; Amazon Environmental Research Institute et al. 2020). In the United States, wildfires in the western states, especially along the coast, are among the worst ever recorded. California is currently enduring its worst fire season on modern record, with close to 1.7 million hectares burning and thirty-one fatalities. Evacuations have been complicated by an historic heat wave and the ongoing pandemic. Officials blame the effects of climate change and poor forest management for the fires' intensity (Cal Fire 2020). In addition to unprecedented droughts, floods, and hurricanes, these fires are among only a few examples of the intensifying natural disasters the world has seen this year.

If the environmental disasters weren't enough, fascism is on the rise globally. Fascism is characterized by authoritarian politics based on leaders who create a cult of personality, brook no dissent, dismiss scientific rationality and evidence, expand policing and surveillance, incite fierce nationalism, proliferate myths, and use perceived victimization of the majority against minorities to allow and promote violence against minorities. Evidence of fascism's recent global ascendency abounds: including Brazil's President Bolsonaro, who has endorsed violence and killing as legitimate political actions; the United States' President Trump, who has emboldened white supremacists and neo-Nazis; Germany's rise of the Alternative far-right political party; and India's Prime Minister Modi, who is a member of a Hindu nationalist organization that engages in a Hindu religious conversion program and has alleged ties to anti-Muslim violence, among many other examples (European Economic and Social Committee 2018). Each of these aforementioned leaders have rolled back environmental protections in their respective countries and openly reject the scientific consensus on climate change. Moreover, they stifle alternative, coalitional, and green political imaginaries essential for change and human survival by polarizing people within their nations through divisive means, such as breeding distrust, fear, and hate of immigrants and racial and religious minorities. However, tyrannical figures also foment antifascist revolutionary formations.

With authoritarian, forceful, and oppressive control comes political resistance—insurgency and counterinsurgency, perpetuating cycles of violence. In the United States, uprisings continue over the state-perpetrated killings and policing of Black, Indigenous, and Latinx communities. Specifically, mounting tensions between police and Black people recently came to a head after centuries of patrolling, police killing, and targeting of Black people as well as their inordinate imprisonment since the 1970s—all preceding the Black Lives Matter protests (Alexander 2010).

In terms of immigration, considering the border wall and the atrocities committed by U.S. Immigration and Customs Enforcement (ICE) officials within, and outside of, detention centers—such as children separated from their families (American Civil Liberties Union [ACLU] 2020), the preventable deaths of ill detainees (ICE 2020; Nowrasteh 2020), human trafficking (Pierce 2015), sexual abuse (López and Park 2018), and nonconsensual hysterectomies performed on detained women in ICE facilities (Treisman 2020)—I shudder to think of what the nation's response will be to the 78 million people projected to immigrate to the United States by 2065 (Pew Research Center 2015). By 2050, 17 million people from Latin America are projected to migrate as a result of climate change (World Bank 2018). Most of them will migrate within their own countries. However, many of them will attempt to find respite from rising temperatures, droughts, and floods by migrating north to the United States. Internationally, climate migrants are not considered refugees, and countries are not obligated to take them in. Without a climate refugee policy in the United States, and with intensifying xenophobia and nationalism, border conflicts will worsen. As an example of this, the US military speaks of climate migrants as potential national security threats and warns:

> Among the future trends that will impact our national security is climate change. Rising global temperatures, changing precipitation patterns, climbing sea levels, and more extreme weather events will intensify the

> challenges of global instability, hunger, poverty, and conflict. They will likely lead to food and water shortages, pandemic disease, disputes over refugees and resources, and destruction by natural disasters in regions across the globe. . . . In our defense strategy, we refer to climate change as a "threat multiplier" because it has the potential to exacerbate many of the challenges we are dealing with today. . . . We are already beginning to see some of these impacts. (U.S. Department of Defense 2014, 1)

Treating indigent human beings as a threat is dehumanizing and puts them in danger of more deprivation and violence. In addition to violating human rights, treating immigrants or minorities as political threats creates unnecessary conflict. Within the United States, internal climate migration is expected to total 13 million people by the end of the century (Hauer 2017). Globally, the United Nations (UN 2016) predicts that there will be 200 million (with reports of up to one billion) climate migrants by 2050 (Milan, Oakes, and Campbell 2016). If the rise of fascism, nationalism, and xenophobia continue unabated, more border and resource wars are immanent.

From where I sit, things look apocalyptic. Despite the ominous connotation of the word apocalypse, which in its modern form means "the end of the world as we know it," the word has a hopeful etymology. From its Latin and Greek roots, the word means a revelation or to reveal, disclose, or uncover. In its Judeo-Christian form, it is a prophetic revelation, whereby a cataclysmic event allows the forces of good to permanently triumph over evil. I choose to interpret this apocalyptic moment as a reckoning and unveiling, a moment where a light is shed over the contradictions and the unjust and unsustainable practices bolstering our current social structures. This revelation, as horrific as current conditions are, presents an opportunity. As the old structures crack and break under the pressure of repeated catastrophes, visionary socioecological modes of living already underway present a multiplicity of alternative paths. The challenge for us all is to be bold and imaginative and to work within our communities to build upon these alternative models. Our collective creativity is needed to scale up regenerative socioecological cultures and practices that will mitigate past harms and keep us adaptive into the future. Again, I urge you, be brave, be bold, be creative. Imagine the impossible and begin living it. The stories in this book will give you some ideas about how and where to begin.

Acknowledgments

As some of my respondents would say, "it takes a village." Of course, they mean this in reference to raising a child (from an African proverb), but I think it applies to writing a book, too. There are many people in my village, separated by space and time, that made this publication possible. I am forever indebted to my gracious participants, the Cuban urban farmers and the US ecovillagers, who shared their homes, stories, time, trust, and wisdom with me. I admire and appreciate all of the important work that they do.

I wish to express sincere gratitude to my mentors, Professors Yvonne Braun, Jim Elliott, Greg McLauchlan, Kari Norgaard, and Stephen Wooten for their assistance with my work in preparation of this manuscript. Of course, the most important mentor of them all is part advisor, family, and friend; I am profoundly grateful to Richard York for supporting me through life's troughs. Richard is second only to my partner, Jacob Eury, who has always appreciated me exactly as I am, idiosyncrasies and all, and who is genuinely proud of me, without competition or envy. Special thanks are due to colleagues old and new, especially Kathryn L. Miller for helping with edits during the dissertation phase of this manuscript (despite volcanic eruptions), and new colleagues who engaged in conversations with me, read various chapters, and provided thoughtful feedback: Lo Presser (who read most of the book!), Michelle Brown, Bobby Jones, David Pellow, Tyler Wall, Kasey Henricks, Jon Shefner, Stephanie Bohon, Meghan Conley, and Louise Seamster. Michelle Calderone's edits and support were invaluable. And special thanks to my hardworking editor, Nancy Doherty, who saw too many older drafts and continued working with me anyway! I also want to extend appreciation to the OUP editors, especially Emily Mackenzie, for helping me through the publication process and the anonymous reviewers. This investigation would not exist if it were not financially supported, in part by the Wasby-Johnson Dissertation Research Award and the Center for the Study of Women and Society Research Grant at the University of Oregon.

I would like to thank my chosen family and friends—Theresa Koford, Sarah and Steve McNeale, Caroline Kaiser, Will Steele-Long, Joy Winbourne, Hannah Holleman, Matt Trimble, Daniel Galvan, Resham Arora, Mary

Laube, Jenny Crowley, Martha Camargo, Julius McGee, Ryan Wishart, Matt Clement, Patrick Greiner, Patrick Grzanka, and Kim Donahey—for conversations, feedback, and good times during the process. I have much appreciation for June Zandona who barely knew me but wanted to work on my documentary idea anyway. Thanks to Albert Wolff and Sam Evans for listening to me talk about the climate crisis while serving me cocktails! I would be remiss if I neglected to mention my dedicated students who work tirelessly to realize ecojustice; they remind me why I bother. Thank you, Emily Landry (Medley), Sam LaVoi, Evora Kreis, Dani Urquieta, Mariam Husain, Kiran Hussaini, Isabella Killius, Sally Ross, Nadya Vera, Bryan Clayborne, and the many motivated young people to come. To the activists, scholars, and scientists unwaveringly doing the thankless work of helping the planet, I see and appreciate you.

I am thankful for the trees who frame my office window: their place-based wisdom reminds me to remain present and calm, and the birds and diversity of life they house and feed bring me endless joy. To the beautiful Earth, I am deeply grateful and indebted to you for providing the diversity of life that inspires me, the ground that supports my body, the food that nourishes and animates me, the energizing and cleansing air, and the water that quenches my thirst and soul. Thank you for being the perfect home despite some unfortunately ungrateful individuals among my species. I am humbled to have been a speck of your dust temporarily imbued with a modicum of your consciousness. I hope to use your gifts to honor you.

Special thanks are in order for my spirited and hilarious cat family, especially my sweet cow-cat who sat with me as I wrote most of this book, and to the memories of my animal friends of the past, a turtle and dog, who taught me that nonhuman animals feel and have agency. To my family, thank you, Noli Ergas, for loving and honoring the Earth as much as I do. Thank you, aunt Lynn, for showing me that women are capable and competent. And to Angela and Enrique, thank you for showing me kindness; I wish I had gotten the chance to know you better. I especially appreciate my mother, Rosa Ergas, for always believing in me.

Innumerable teachers of my past have inspired my curiosity, and because I cannot remember them all by name, I thank anyone who teaches. Indeed, I'm grateful to anyone who does the hard work of caring for their fellow community members; whether as nurses, farmers, doctors, nannies, ecojustice activists, or foresters, thank you for doing care-work. Care-work is the work of the future. If I've left anyone out, please know that I appreciate you. And to anyone who reads this book, thank you for considering my words, and please show your love to the Earth.

Introduction

Building Socioecological Community

> We have learned that it is not enough to define utopia, nor is it enough to fight against the reactionary forces. One must build it here and now, brick by brick, patiently but steadily, until we can make the old dreams a reality: that there will be bread for all, freedom among citizens, and culture; and to be able to read with respect the word "peace." We sincerely believe that there is no future that is not built in the present.
>
> —Juan Manuel Sánchez Gordillo, mayor of Marinaleda, Spain[1]

A few years ago, when this book was just a few thin chapters and jotted notes, I lived in Providence, Rhode Island. I regularly commuted on the I-95, an elevated highway by the harbor. One afternoon, as I inched along in rush hour traffic, I turned my gaze toward the industrial port that sits a couple thousand feet below. A thick haze surrounds this New England highway and the nearby port industries, emitting smells of sulfur, exhaust, and other unidentifiable fumes. What could be a beautiful view of a quiet bay is instead an industrial row of barges and freight, rusted scrap metal heaps, and a coal pit. On this afternoon, the blare of ambulance sirens felt like the final assault on my senses, and my thoughts escaped to the winter five years earlier when I worked on an urban farm in Havana, Cuba.

There, from the second-floor window of my apartment in the northeastern edge of the city, I could see a sweep of green: the farm was just a short walk across the street. Although there was some traffic, the farm's twenty-seven acres of trees, crops, and ornamental plants cleaned exhaust from the air and provided habitat for hundreds of bird species. The bustle began early as neighbors visited with each other on their walks to work or taking their children to school. Most items that residents needed were within walking

Surviving Collapse. Christina Ergas, Oxford University Press. © Oxford University Press 2021.
DOI: 10.1093/oso/9780197544099.003.0001

distance. Connected to the neighborhood street was the farm's vending area, where residents stood in line to purchase papaya, mixed-spice packets, and cassava, or, on a hot day, freshly squeezed sugarcane juice.

In Havana, my workday began at eight o'clock in the morning in the plant nursery. Four other workers and I planted seeds and cared for seedlings. The pace was moderate and steady, with long breaks for morning snack, lunch, and afternoon banter. All my work took place outside under the shade of neem trees or thatched roofs made from palm branches. At the height of the sun's full beam, the shade provided a respite where we could sit and enjoy a cool breeze. People sang as they worked and listened to music from radios at their workstations. The red limestone soil coated our boots and clothes with a rust-colored film. The farm smelled of wet clay, leafy greens, and flowers.

It was a far cry from this New England view. The last two years I lived in Providence, residents began a community effort to stop the siting of a natural gas liquefaction (LNG) facility at the main port, only three miles from where I worked and five miles from where I lived. The purpose of the facility is to liquefy natural gas in order to condense it, thus making storage and transport more convenient; it also will serve as the main distributor of LNG to the Northeast region. The port already contained a variety of hazards, including a chemical plant, a petroleum terminal, a cement plant, and coal, propane, and natural gas storage facilities, to name a few. Another such facility would increase the risk of potential disasters. Indeed, complications caused by sea level rise, a hurricane, or a terrorist attack could trigger a cascade of explosions, unleashing a toxic cocktail and sending a chemical plume into the air over a fourteen-mile radius, which would completely envelop the city. The resulting explosions would kill the closest people on impact and injure many others further away. Unsurprisingly, those who live nearest the port are the city's poorest and most ethnically diverse and have the least political sway. As of this writing, the Federal Energy Regulatory Commission has approved the natural gas liquefaction facility. The state Department of Environmental Management has reviewed the plans and issued the final permits, and the facility is in the process of being built.

The US fossil fuel-driven economy thrusts social and ecological costs onto many communities similar to the residents of South Providence, RI. These sacrifice zones appear in stark contrast to urban agriculture and reforestation efforts I observed in Havana, Cuba. I don't mean to make light of the fact that Cubans suffer from a lack of material resources, including serious energy deprivation, especially in fossil fuels. However, through their ingenuity and development of community resources, Cubans have managed to inspire hope worldwide despite their energy shortages. I argue that Cubans are able to thrive under austere conditions because they have prioritized social and community resources, and their experience offers valuable insights for every community facing the impacts of climate change—primarily caused by the burning of fossil fuels.

I.1. Running on Fumes

Ever-increasing amounts of fossil-fuel energy feed the globalized growth economy, enabling society-wide addictions to excess consumption. This unabated economic expansion is driving us toward self-destruction as well as to the destruction of most other species, a phenomenon scientists have termed the sixth mass extinction (Ceballos, Ehrlich, and Dirzo 2017). For decades, international scientific associations, such as the World Wildlife Fund (WWF) and the Intergovernmental Panel on Climate Change (IPCC), have been warning inhabitants of the Global North, especially its most affluent, to change their relationships with the environment and with each other.

The current global economic prescription for our converging environmental and social crises is more sustainable development, or, as the most widely cited Brundtland Commission (UN 1987) defines it, development that "meets the needs of the present without compromising the ability of future generations to meet their own needs." As the term "development" implies, this consists of continued economic growth through new "green" technologies and the expansion of global markets. In terms of agriculture, sustainable development includes corporations genetically engineering seed to increase herbicide and drought resistance in monocultural harvests rather than changing production processes to allow for biodiversity as well as carbon and water retention.

As this book will reveal, Western production and consumption of new technologies at ever-increasing rates is in part what created this situation in the first place. Instead, we should be questioning technological production processes, whether there is need for so many new gadgets, and prioritizing technological production that serves community resilience. We are reaching the limits of Western societal hubris. Business-as-usual is not going to pull us out of this mess. In this book, I identify the roots of socioecological problems, and I argue that changing social relations, or systems of power and inequity, through more *socially sustainable* practices, is what is needed for greater societal resilience. The cases that I describe offer lessons on how people adapt to pressing ecological problems; they do so by fostering community and reclaiming self-sufficiency. Insights from these communities can help us imagine different stories of survival. I maintain that with a solid vision, we can move beyond debilitating fear and denial toward a just transition to a new economy. This new economy is based on social equity and environmental regeneration, or a radical sustainability that is at once socially and ecologically transformative—dismantling hierarchies toward total liberation—and regenerative—healing and restoring the health of people and the planet.

One way to wean nations off of fossil fuel addictions is through food sovereignty, or the relocalization of food production and consumption. According to the UN Food and Agriculture Organization (2017), modern food production, including "input manufacturing, production, processing, transportation, marketing, and consumption . . . accounts for . . . approximately 30 percent of global energy consumption" and produces more than 20 percent (or, depending on the source, from 24 percent up to one-third!) of global greenhouse gas emissions (Gilbert 2012; Environmental Protection Agency [EPA] 2017). Specifically, in the United States, agriculture accounts for 80 percent of nitrous oxide and 35 percent of methane emissions—two powerful greenhouse gases—and, worldwide, up to 25 percent of the carbon dioxide emissions (Shiva 2016; EPA 2017). The US Department of Agriculture (USDA; 2017) reports that in the United States, fossil fuel-derived energy makes up 93 percent of the energy used in agricultural production. The type of fossil-fuel energy depends on the region. In most of New England, such as Rhode Island, 98 percent of electricity comes from natural gas, whereas other regions, like West Virginia, rely mainly on coal.

Food sovereignty produces food *for the local community by the community*, which allows members to decide on the most locally desirable and suitable foods and best practices. It reduces food transit miles and the need for refrigeration, and it allows for regional variation and diversity. Regenerative organic agricultural practices that utilize crop rotation, no-tillage, and composting also have the potential to sequester, or remove from the atmosphere, up to 15 percent or more of global carbon emissions (Lal 2004; Gattinger et al. 2012).

This book explores in depth two cases of living arrangements and food alternatives that differ from the dominant global, United Nations vision of sustainable development. Urban agriculture in Havana, Cuba and in an urban ecovillage in the Pacific Northwest are examples of experimentation in more holistic sustainabilities that encompass food self-sufficiency, participatory democracy, and equitable social relations. They differ in context, as the case of Cuban urban agriculture is part of a top-down government plan, while ecovillages spring up in the United States independently and in opposition to government and corporate policies. These cases illustrate how cultures and individuals respond differently to their environments, working within the limits of scarce resources and in relation to their unique ecosystems. In ecosystems, diversity is important because biodiverse systems are more resilient to rapid environmental change—our current global reality (Jenkins 2003). From these two cases, and many more like them, we can learn that building an economy that cooperatively and symbiotically works with nature—rather than the hegemonic, competition-driven, domination and control orientation toward nature—is the way out of converging environmental crises. These cases demonstrate some, though by no means all, of the ways to begin healing individual and societal rifts from nature. We can expand on these lessons to move toward a radical sustainability that is at once a socioecologically regenerative and transformative, rather than "sustainable," way of life.

I.2. Anthropocene (or Capitalocene?)

We live in a time that is heavily marked by human activity: every space on this planet has been affected by people. As a result, scientists are calling this epoch the Anthropocene, after the root Greek words for human and new (Zalasiewicz et al. 2008; Stromberg 2013; Barnosky et al. 2014; Steffen, Crutzen, McNeill 2007; Steffen et al. 2015). Crutzen, the Nobel-prize-winning atmospheric chemist who popularized the term, defines it as "human dominance of biological, chemical, and geological processes on Earth" beginning in 1800 (Crutzen and Schwägerl 2011). Others characterize such dominance as the Capitalocene to reflect the specific burden that capitalism as a global economic system has placed on the planet (Haraway 2015). What is clear is that our converging environmental and social crises are social in origin. The evidence is mounting; scientists are coming to see that the way Western societies live, especially the more affluent among them, wreaks havoc on ecosystems. What we need are revolutionary changes in social and economic arrangements, a radical sustainability that is at once regenerative and transformative. This book is about what making these changes might entail.

The global economy is characterized by perverse incentives. Its operating logic situates cyclical ecologies in a linear, factory-like, output-oriented system called commodity chains. It is set to promote unyielding growth; it encourages cut-throat competition, orienting corporations to focus on profits, ignoring the external costs or so-called negative externalities. These externalities are the social and environmental harms—such as toxic waste and its associated community health problems—that corporations comfortably avoid paying for. The value of nonrenewable natural resources is systematically underpriced, because the needs of future generations are not accounted for in a short-term profit structure (Wright 2010). In this growth-driven economy, consumerism is promoted through market-expansion around the globe; privatization and commercialization of common resources; marketing and advertising; as well as the planned and perceived obsolescence of goods (goods intentionally made with limited lifespans) to keep consumers buying things that they don't need (Schor 1998; Dawson 2003; Wright 2010). The consumption of goods implicates the chain of resource extraction, manufacturing, packaging, selling, and waste production common in the global economic system (Bhada-Tata and Hoornweg 2012). To accomplish this, multinational corporations—which are facilitated by international trade agreements negotiated asymmetrically between hegemonic nation states and less developed countries—market rugged individualism, promote excessive work hours, and buy up once communal land and parcel out private property, thereby dividing and conquering peoples and lands. The tactics in effect atomize, exploit, dominate, and control.

Through the charade of sustainable development, this exploitative way of life is exported to countries, many of which are former colonies of Global North nations, around the globe. With the goal of economic development, multinational corporations and development agencies promote Western values that homogenize cultures and economies, coaxing them to participate in global economic activity. Thus, impoverished people around the world are marketed, and may welcome, excessive consumer lifestyles like those in the West. Why shouldn't they want the comfort and convenience that Westerners enjoy? However, rather than enjoying the spoils, the poor among the poorest and richest countries are deceived, their labor exploited, and their natural resources and wealth taken from their lands to benefit some at the expense of many others (Sekerci and Petrovskii 2015). In order to combat the converging environmental and social crises that scientists are calling anthropogenic, or human caused, in origin, Western societies will need to radically alter their lifestyles in accordance with ecological limits and social equity.

In this book I argue that under the above paradigm, newly created, "green" technologies fall short of creating more sustainable environments and social relations: only affluent people can afford them at first, and when they become affordable, then many people may purchase the technology, which requires more resource exploitation and production and does not necessarily phase out less efficient technologies. Specifically, in the following chapters, I detail how those green technologies miss the mark when they are scaled up in the current global economic context, as even industrial organic agriculture does not fulfill the promise of cleaner streams and reduced greenhouse gas (GHG) emissions. Therefore, "mainstreaming" green in the current context is not the answer. Instead, the cases featured in this book offer hopeful and doable examples of sustainability that offer paradigmatically different approaches to combating socioecological disintegration. Volumes of books exist that expose greenwashing—marketing misinformation designed to give a product an environmentally responsible public image—and how market mechanisms and green capitalism have failed us (Rogers 2010a, 2010b; Holleman 2012; Zehner 2012; McGee 2015). This book explores elegant, low-tech alternatives that mitigate socioecological harm and are adaptive to already changing environments.

As a citizen of the United States, I think it's fitting to look at my country first. The United States accounts for less than 5 percent of the world's population, but consumes about 25–30 percent of the world's energy and produces about a quarter to a third of the world's waste (Ponting 2007; EPA 2013; Hoornweg, Bhada-Tata, and Kennedy 2013; World Watch Institute 2016). Further, US citizens emit about sixteen metric tons of carbon dioxide per capita yearly, as compared to the five metric tons people emit worldwide (World Bank 2021). The United States is not alone in excessive consumption and waste production; it's just one of the top contributors. Member countries of the Organization for Economic Cooperation and Development, including the most developed economies of the world and a few newly emerging economies, are among the highest waste generators as well[2] (Hoornweg, Bhada-Tata, and Kennedy 2013).

The amount of US-based waste and emissions should not come as a surprise. Individual consumption is a vital part of the economy, and household consumption amounts to about 70 percent of the US gross domestic product (GDP); worldwide the average is about 60 percent (World Bank 2017). Marketing firms and advertisers spend a trillion dollars yearly to ensure that US households continue excessive consumption habits (Dawson 2003; E-Marketer 2014). Consumer culture, with its treadmill of long work weeks, car commutes, excessive product packaging, energy consumption, waste generation, and excessive meat consumption (Pollan 2006), negatively affects the environment and, despite the advertisements, it doesn't even make people happy!

In fact, depression is on the rise globally, as are mental health conditions related to environmental stressors, like eco-anxiety (Layard 2005; Clayton et al. 2017; World Health Organization [WHO] 2017). Working more to consume more has diminishing returns. Money and consumption only increase happiness and well-being up to a certain point. As researchers have observed, "once a person has a comfortable standard of living, increased income and consumption do not lead to increased life satisfaction and happiness" (Wright 2010, 68). What gives people meaning and makes them happy is connecting with other people and the environment, engaging in creative and interesting work and activities, and participating in community (Layard 2005; Wright 2010; Clayton et al. 2017). Working more to buy more only serves to take time away from what is truly important to people: their relationships.

Problems on every societal level—from the global (material flows and international markets), national (regulations, policy implementation, subsidies and incentive programs, and planned development), local (re-localizing food production and human-scale technologies), all the way down to the individual (processing emotions and changing ideas and behaviors)—are interconnected and contribute to the converging crises. Therefore, holistic systemic solutions are necessary to combat these problems.

I.3. The State of the Biosphere

Climate change is arguably the greatest environmental threat that humanity has ever confronted. It is global in scale, affects all other planetary systems, and is marked by increasing global average temperatures, rising sea levels, and more extreme weather patterns. Ninety-seven percent of scientific articles on climate change agree that it is happening and that human activities are largely responsible for it (Crowley 2000; Intergovernmental Panel on Climate

Change [IPCC] 2013; Stern and Kaufmann 2014; Smith 2015; Zalasiewicz and Williams 2015; Cook et al. 2016). Climate change is accompanied by other interrelated environmental problems, including ocean acidification (Pelejero, Calvo, and Hoegh-Guldberg 2010; Longo, Clausen, and Clark 2015; WWF 2015), potable water decline, dustbowl-ification (Romm 2011; Holleman 2017), and mass species extinction (Pimm and Raven 2000; Pimm et al. 1995, 2014; Lydeard et al. 2004; UN 2005; Burkhead 2012; Cardinale et al. 2012; WWF 2014, 2015; Régniera et al. 2015). Each problem alone can have grave consequences for human societies, such as resource wars, plagues, and mass migrations. But, of course, these problems are interconnected, and their convergence is catastrophic. To make matters worse, economic trends leading to more consumption and waste have been accelerating since the 1950s (Steffen, Crutzen, McNeill 2007; Steffen et al. 2015). Scientists have termed this the Great Acceleration (Steffen, Crutzen, McNeill 2007).

Anthropogenic climate change is largely driven by rising atmospheric concentrations of greenhouse gases—primarily carbon dioxide, but also methane and nitrous oxide—that trap the sun's heat on Earth rather than allowing it to escape into space. The burning of fossil fuels is the main culprit for the steady and rapid rise in concentrations of atmospheric CO_2. If global emissions continue to rise, scientists predict a likely four-degree (or more) Celsius rise in average global temperatures by the end of the century, a temperature that few living things could adapt to in such a short time (IPCC 2013; World Bank 2012; Sherwood and Huber 2010). Indeed, temperature increases of even less than that are associated with interconnected environmental problems that exacerbate extreme weather events, such as longer or mega-droughts (lasting a decade or more) that affect food production, more dangerous hurricanes and tornadoes, monsoons, and increased flooding; glacial melt that causes sea level rise, inundating coastal cities and covering island nations; surging human wars and conflicts over resources, mass migrations, malnutrition, deaths due to heatwaves, and pandemic plagues (Hsiang, Burke, and Miguel 2013); expanding regions experiencing exacerbated water and food scarcity; irreversible biodiversity loss; and ocean acidification and marine-life collapse (World Bank 2012; Sekerci and Petrovskii 2015).

Intensive industrial agriculture practices not only contribute heavily to greenhouse gas emissions, because of the use of fossil fuel-powered heavy machinery, food transport, refrigeration, processing, and excessive use of nitrogen fertilizers (often emitted as nitrous oxide), but they also degrade soils and cause soil erosion. Healthy soils are second to our oceans as the best sources of carbon storage. Humans release carbon by clear-cutting forests for large-scale agriculture and degrade healthy soils through monoculture, and these processes only serve to exacerbate climate change. With severe weather changes, an increasing number of regions will experience severe droughts or excessive rains, either of which will make agriculture and soil nonviable (Gattinger et al. 2012; Holleman 2017). There are many other examples of such eco-systemic problems.[3]

Scientists have formulated potential roadmaps for keeping us at about a 1.5-degree Celsius global temperature rise, an increase in temperature to which they believe humans can adapt. But if we do not begin down this path now—curbing emissions by half every decade until ultimately cutting emissions to net zero by mid-century—it will soon be too late (Rockström et al. 2017; IPCC 2018). Some question whether we have the political will to make the necessary changes (Victor 2009). A US Trump presidency certainly did not bode well for climate legislation (Gore et al 2016; Davenport and Rubin 2017; Merica 2017). Trump-era policies—such as disbanding the federal climate advisory panel, appointing an EPA administrator who actively fought environmental regulation, and suspending environmental risk research (Greshko et al. 2017)—will only serve to propel us further into a climate-nightmare future.

What can we do about these global problems caused by economic and social arrangements? As individuals? As societies? At each international climate summit, most recently in Madrid, nations negotiate agreements that don't go far enough to mitigate climate change (Cama 2015; Davenport 2015). Of course, there are examples of environmental problems that we have combatted at larger scales, such as the universal ratification of the Montreal Protocol that required nations to phase out production of ozone-depleting substances (United Nations Environment Programme 2017); reforestation in some places (Clement, Ergas, and Greiner 2015); and the US Endangered Species Act that has saved numerous species. However, relatively little action is being dedicated to alleviating climate change.

At the time of this writing, the end of the twenty-first century is about eighty years away. If we do nothing more to mitigate climate change, we will likely see a two-degree Celsius or more rise in average global temperatures in the next twenty to thirty years. Scientists predict that this will still change much of life on the planet. We have already seen the hottest years on record every year from 2014 through 2020 (NASA 2021; National Oceanic and Atmospheric Association [NOAA] 2019). The first two decades of the twenty-first century have seen twenty of the warmest years on record (NASA 2021). We have already experienced extreme heat and weather events the world over. There are already parts of the United States, specifically Louisiana's southern boot, becoming inundated by water and forcing a whole community to migrate (Marshall 2014). We are already experiencing mass-species extinction (Broswimmer 2002) and coral-reef bleaching (NOAA 2017). The South Sudanese, Somalians, and Syrians are already enduring major droughts and famines that are exacerbating resource wars (United Nations Environment Programme 2007; Biello 2009; Hsiang, Burke, and Miguel 2013; Gleick 2014; Maystadt and Ecker 2014; Beaubien 2017). Temperature rise means that such occurrences will only get worse and spread the world over. In recent years, as the effects of climate change have escalated, some politicians and news agencies in the United States have made blatant efforts to ignore or deny climate change altogether (Cillizza 2017; Public Citizen 2017), which simply compounds an already bad situation.

Perhaps we should pause for a moment. This is indeed bad news, and it is a lot to take in.

Some of us cope by denying the existence of these problems. Some of us are aware of the issues but completely dissociate, watching ourselves contribute to the problem as if we are watching someone else. Some fight with others in social movements; some escape and build their own utopian futures; some are forced to work themselves to death and cannot participate in the conversation, while others profit from the whole mess. Inaction is not an option. Now more than ever we must examine the viability of consumer lifestyles and living arrangements and consider paths different from the globalized and homogenizing forces of sustainable development, which are modeled on a Global North monoculture.

Now is the time to take the work of sociologists seriously, because our insights could help save humanity. Human societies and the social relations within them are causing these converging environmental crises. If we are to stop them, we have to change the way people interact and live. Sociologists should be doing the work of not only illuminating how human societies have created these unsustainable conditions, but also theorizing about and empirically testing projects that model potential solutions. People have been tinkering with alternatives for decades and, in some cases, throughout human history. The technologies and models abound. What are needed are systematic evaluations of these models and the social and political will to scale up radically different social relations.

The problems are interconnected, and examples of holistic, transformative, and regenerative solutions exist. Our immediate attention to solutions is necessary.

I.4. Why Study "Real Utopias"

Sociology as a discipline spends a great deal of time focusing on social problems, dissecting and disentangling injustices that are maintained through our social interactions and institutions. As a discipline, it does less to explore viable alternatives or develop a vision of what we should be striving for. While diagnosing and critiquing social problems are necessary and important, so too is developing a theory for and empirically testing potential solutions to these problems.

Social psychological research suggests that individuals need stories about solutions so they don't suffer compassion fatigue and fall into a state of denial about pressing environmental problems. Giving people more facts about environmental crises does little to empower them; it actually overwhelms them with negative emotions like fear and guilt and renders them debilitated. On the other hand, informing people about existing projects like urban farming and ecovillages empowers them to act in the face of environmental uncertainties while grounding them in the everyday challenges to social change. When people know that there are things that they can do, they develop a sense of self-efficacy that in turn makes them feel more concerned about these problems (Krosnick et al. 2006; Kellstedt, Zahran, and Vedlitz 2008; Norgaard 2009, 2011). That concern can more easily translate into action when people have exposure to viable alternatives. That's what this book seeks to do: explore existing alternatives.

Not long ago I discovered firsthand how valuable this work of envisioning alternatives is. On the last day of both my Social Movements and Gender and Society courses, I asked my students how they imagined societies with comprehensive participatory democracy and gender equity. They gave answers like "people could work one job instead of two to make ends meet," and "women could work the high-powered jobs and their husbands could work the lesser-paid care jobs." Their answers revealed that they had a very limited ability to imagine societies different from their own. I felt I had either failed them and not taught them well that semester, or that they had no experience with thinking outside of what they knew from their own personal experiences. I preferred the latter, more generous interpretation. I know that before I'd seen one, I never could have imagined something like Havana's urban farms. Perhaps this was the problem my students were encountering. They had never seen people experiment with true democracy or equity. Although these are anecdotes from students in only two courses, it made me think more about envisioning "real utopias" as a timely and important task (Wright 2010).

Erik Olin Wright (2010) suggests that we investigate the feasibility of radically different kinds of institutions and social relations that could potentially advance the democratic egalitarian goals historically associated with the idea of socialism. To this I would add, *and that benefit the natural environment*. Wright (2010) and Litfin (2014) offer ways to evaluate the "success" of the alternative projects. Litfin (2014, 13–14) specifically suggests we measure longevity of the organizations, size, resource consumption and waste, economic prosperity, ripple effect, cohesiveness, embodied vision, as well as happiness and satisfaction. Wright (2010) suggests we measure the viability, desirability, and achievability of such projects. I'd also add measures of ecological, carbon, and water footprints as well as social equity. The cases I highlight have passed the tests set forth by Litfin and Wright and have proven malleable enough to be viable for people in different environments and cultures.

The cases that I explore are not meant as prescriptive paths toward sustainability; instead these examples of alternatives are meant to whet the appetite of those craving a different way. Wright (2010) cautions against attempting to implement utopian fantasies with no prior trial or assessment. He warns of catastrophic unintended consequences. Instead he advocates for "utopian ideals that are grounded in the real potentials of humanity, utopian destinations that have accessible waystations, utopian designs of institutions that can inform our practical tasks of navigating a world of imperfect conditions for social change" (6).

I argue similarly that we should consider utopian ideals something that we should constantly strive for rather than something we can achieve, using them as orienting visions in our ongoing and honest critique of best practices. We shouldn't hold up model sites as the end game, but rather a stop on the way toward something more adaptive to the contemporary needs of people and the natural environment. I appreciate Litfin's (2014) description of ecovillages not as utopias,[4] but rather as living laboratories where people are experimenting with alternative pathways, tinkering with self-sufficiency and participatory democracy.

Inaction happens for a variety of reasons and at multiple levels. At the individual level, compassion fatigue, hopelessness, and helplessness trigger denialism and inaction. As I previously noted, people need examples of viable and socially desirable alternatives to consumer lifestyles in order to believe that another way is possible, to be capable of envisioning a different society, and to act to change things. We need to deliberately think through and practice the process of change in order for it to happen in a way that we intend. We have to visualize and begin living that change in order to understand its flaws and unintended consequences, so that we can work out the kinks, major inconsistencies, and failings, in order to stand strong during the major catastrophes that threaten to break us. Some questions to guide the visioning process include: How can we organize to create socially just and sustainable environments where we live? What are people already doing that we can tailor to fit our local culture and bioregion? How can we create an economy based on ecological and social transformation and regeneration?

We need resilient systems in place so that we can weather the major storms ahead.

I.5. Experiments in Surviving Collapse

Some people worry that leading a more environmentally sustainable life means leading a life deprived of comforts. This was the meaning of the title of Al Gore's (2006) documentary about climate change, *An Inconvenient Truth*. Gore ends by proposing consumer solutions to reduce our intake of energy and curb our individual emissions. Certainly, there are items and practices that some individuals should learn to live without, such as regular air travel. However, what the cases I discuss in my book demonstrate is that the few things we may need to give up can be offset by the many benefits we stand to gain, such as a supportive community, self-efficacy, a closer relationship with nature, and better mental and physical health. But if we, as communities, do not plan for what may come, prefiguring the socially just and environmentally regenerative future we'd like to live, then our future may look like any number of dystopian novels, for which the US military is actually preparing and that some countries are already enduring (Center for Climate and Security 2017). This is a future world where displacement, war, famine, and plagues claim the lives of many, and the rest are forced to fight for their survival. However, many of us can still choose the path we take, and we can learn from our living laboratories. The time to think through a just transition, or a socially equitable and ecologically based economy, is now.

Cuba, considered a poor or less-developed nation by Western standards, has valuable lessons to impart on how to take on converging economic and environmental crises, and some living eco-experiments in the United States illustrate how to implement these lessons in the "developed" world. Cubans faced a national-scale economic and environmental collapse in the 1990s, and, through careful planning, Cuba arose as a model nation globally by implementing restorative environmental programs and moving toward food sovereignty. They did so while maintaining high levels of human development and happiness and becoming the most sustainable economy in the world (WWF 2014). Given our global-scale converging crises, Cuba's experiments provide valuable insights about how to work cooperatively to build a more resilient world.

However, Cuba isn't the only place experimenting with alternative food production, self-sufficiency, and more equitable social and economic relations. Indeed, we also can take a page from prefigurative social movements, like ecovillages, to see how to generate small-scale local economies, living and social arrangements that mimic ecosystems, and engage in participatory democracy. There are many examples in the United States and around the world of such experimentation happening on a much smaller scale. The thing is, we need examples on all scales, from the local to the national. If a handful of nation states or localized social movements can pull it off, then let's learn from them.

Specifically, in the United States the ecovillage movement is a growing co-housing phenomenon that experiments in self-provisioning, food production, direct democracy, green building, and communal living (Ergas 2013ab). I spent two years learning from and living in an ecovillage in the Pacific Northwest. The first time I saw the ecovillage, I was twenty-four and had recently graduated from a university. I had never heard of an ecovillage, or anything even approaching it, and when I saw it, I was in awe. Homes made of mosaic cob, an earthen material, lined the perimeter of the village. Vegetable gardens, fruit trees, and community gathering spaces filled the center. The space was vibrant, and I felt inspired by this creative living arrangement that captured some of the organic fluidity of a thriving and diverse ecosystem. There were bees and their hives, rabbits, chickens, geese, stinging nettle, plum, apple, and pear trees, corn, kale, and tomatoes all living together with people.

The ecovillage was such a huge departure from the cookie-cutter suburbia I grew up in, one typical of much of middle America. Seeing something so different opened me up to the range of possible living arrangements. Growing up, I took it for granted that most of my friends and family didn't know their neighbors; that our homes had manicured grass lawns; that homes were built with wood frames, dry wall, and plastic paneling. But in fact this living arrangement is a relatively new phenomenon (since the mid-twentieth century) and evidence indicates that it is alienating and isolating (Slater 1970; Schor 1998; Putnam 2000; Layard 2005), as well as potentially damaging to mental and physical health (Oliver 2003; Sturm and Cohen 2004). I know I'm not alone in taking my way of life for granted, which is why exposure to alternatives is so important to the future health of our communities.

Some of the most significant lessons from Cuban urban agriculture and the ecovillage cannot be scaled up in the dominant individualist consumer society (Schor 1998; Dawson 2003; Wright 2010). The ecovillage model is about downshifting—or working less, earning less, and consuming less and more deliberately (Schor 1998)—as much as it is about communally sharing space, knowledge, tools, and food (Litfin 2014). The ecovillage highlights the opportunities we all have right now. We can engage our friends and neighbors to begin living the change we need immediately. We can limit the need for resources and energy just by mimicking ecosystems' natural flows, as permaculture explains. Cuba's local and organic agricultural system demonstrates that we don't need high-tech solutions to live a life with high human well-being (as defined by the UN and explored in depth in chapter 3) and sustainability; that we can do this with minimal resources; and that we have everything we need right now to make the change. The Cuba case also exhibits that high-tech solutions may not be ideal solutions at all and may in fact be undesirable in a resource-limited world. Cuba's struggles reveal what the future may hold for other nations. Scaling up these insights could simply mean culturally appropriate versions of these models that match particular biospheres and preferences, norms, and values, or it could mean scaling up the values of equity and process-oriented systems thinking that these models illustrate.

Through visiting these model sustainability sites, I realized that other ways of living and working are possible and desirable, ways that include more community interaction, connections with other species of plants and animals, and creative and fulfilling work. These sites overwhelmed me with hope and inspiration. I hope that, for anyone reading this book, it inspires a similar sense of curiosity that motivates you to create a better future as well.

I.6. A Challenge for the Future

Despite the dire environmental predictions I have outlined above, my book is intended to offer hope. Change can happen now for anyone willing to organize with others, rather than waiting for governments to make sweeping changes. Changes can begin to take place in our own backyards, in our neighborhoods, in our cities.

In his book *Collapse*, Jared Diamond (2005) argues:

> because we are rapidly advancing along this non-sustainable course, the world's environmental problems will get resolved, in one way or another, within the lifetimes of the children and young adults alive today. The only question is whether they will become resolved in pleasant ways of our own choice, or in unpleasant ways not of our choice, such as warfare, genocide, starvation, disease epidemics, and collapses of societies (498).

I propose, like Diamond, that many of us have a choice. These changes will happen; and we can be prepared and change things on our own terms, or we can wait until change is forced upon us. The choice is ours. But despite what individualist, bootstrap mythology would have us believe, we cannot achieve resilience alone. We will have to work together and, through our collective efforts, create a larger movement. Collaboration and interdependence are the values that will deliver us from this self-interested race for accumulation no matter the socioecological cost. We have to learn to value the wealth of having healthy, connected communities, rather than ceaseless competition to outperform our neighbors' consumerism.

I know change is challenging, and more challenging for some people than others. It takes time to develop community relationships, to acquire skills and information, to alter the necessary policies and laws in order to accommodate changes, and finally to implement changes. Some changes require using money and precious resources. Some communities offset these costs by sharing tools and resources. But the practices I discuss in this book are relatively cheap in comparison to other so-called green technologies (such as all-electric cars), and they can be done on a human scale, which means most people can learn, use, and fix them. They are also doable almost anywhere: people are doing or using these practices somewhere near you. We can learn from each other by bringing together skill-sharing community groups. We also can learn from people or organizations already well practiced in re-skilling.

The sooner we all begin working together to make these changes and learn these skills, the easier and more affordable it will be, and we will be more adaptable and resilient when we can no longer ignore the environmental problems to come. The time to connect with others and act is now. See it as investment in your future, your retirement plan, an investment in yourself, your survival, and the survival of your posterity. See it as restorative for your community, other species, and the environment.

As the hegemonic global economy flails and falters while breathing its dying gasps, it attempts to take down everything else in its path. But another way is possible and necessary, and that is what this book is about, exploring viable alternatives so that we may survive collapse.

What follows are summaries of the chapters in this book.

I.7. What's in the Book

This book is organized into six chapters. Chapter 1, "In the Shadow of Sustainable Development," is heavily theoretical and provides historical context. If you are less interested in the theory and history or prefer to focus on the alternative cases, skip Chapter 1 and begin on chapter 2. If you enjoy reading history and theory, chapter 1 explains the history and sets up the paradox of urbanization. It delves into environmental theories, the social psychology of climate denial, the problem of weak neoliberal sustainable development, and the definition of a radical sustainability—which is at once socially and ecologically transformative and regenerative. Sustainable development, as it has been conceived, is actually a shell game for creating neocolonial dependency in the developing world rather than more sustainable self-sufficient nations. I present my cases as alternatives to sustainable development and as examples of radical sustainability and self-sufficient, autonomous development.

In chapter 2, "Grassroots Sustainability in a Concrete Landscape" I explore the urban ecovillage in the Pacific Northwest, my first example of radical sustainability and autonomous development. Through participant observation and interviews, I present the challenges ecovillagers face attempting sustainable living in a neoliberal context. I also examine the cultural conflicts between sustainability culture and consumer culture and the exclusive, upper-class, white nature of the local food movement in the US context (Molotch 1976; Parr 2009; Alkon and Agyeman 2011). The monetary and time constraints associated with growing local, organic food has largely turned it into an elite phenomenon in the United States, such that it is relegated to those with the disposable income, luxury of time, and education to grow produce for personal consumption. In fact, urban gardening is often cited as a first step to gentrification in urban communities (Guthman 2004; Quastel 2009; Checker 2011).

The third chapter, "Urban Oasis," focuses on an example of radical sustainability in the Cuban context. Cuba's history with resource scarcity exemplifies the myriad struggles that will continue to spread the world over. Cubans were able to transcend their worst problems through top-down governmental policies that subsidized urban agricultural food production. These top-down policies established what are now networks of relatively autonomous, decentralized urban farms that feed surrounding populations. At a Cuban urban farmworkers' cooperative, farmers are residents of the local community who sell their produce at a low cost on site to their neighbors and passersby. By resisting neocolonial forces, Cuba is one of the few countries in the world to obtain high human development and consume a fair share of environmental resources, maintaining a small ecological footprint. I worked on one such farm and lived with a nearby family for several months, and I was able to learn much about the Cuban program's successes as well as the social and environmental barriers it still faces.

The fourth chapter, "Beyond Neoliberalism: The Promise of a Communitarian Story," explains the cultural stories and values that bolster the neoliberal paradigm, one that shapes exploitative socioecological relations. It argues that ideas have consequences and details the history of Western thought—such as Descartes' hierarchical dualisms and the social sciences' profound misunderstanding of Darwin's theory of evolution—that brought about extreme individualization, inequality, and fierce competition. These stories and values promote the ideas that humans have moral dominion over nature as man has dominion over woman. This is a world view that justifies humans' exploitation of other species and the environment as well as social inequity. It is these codified stories and values that perpetuate humans' acts of harm against others and the planet. I further discuss how and why economic context matters in shaping paths of resistance and co-opting alternative and green technologies. I explain the need to scale up socioecological values first in order to cultivate the underlying framework for a new environmentally just economic paradigm based on a radical sustainability.

The fifth chapter, "Scaling Up the Values Themselves: Comparing Two Urban Socioecological Experiments," I develop a radical sustainability framework by examining the socioecological values in sustainability experiments that exist in different political-economic contexts. I ask: What are the environmental values and stories that each of my cases demonstrate? I make the case for a paradigm shift from human's war with nature to human's collaboration with nature to regenerate a thriving biosphere. This means shifting Western culture away from one of atomized, competitive, self-interested individual consumers who use technology to dominate nature. It also means shifting to a culture where cooperation with the socio-natural world, symbiotic relationships, "plentitude" (or sufficiency), and local self-reliance, as well as physical, emotional, and spiritual well-being for all are valued. In addition, it means creating an economy built on social justice and environmental regeneration.

In the concluding chapter, "There Is No Future That Is Not Built in the Present," I sum up the lessons learned from my examples. This final chapter offers practical solutions toward radical sustainability for communities unable to build ecovillages or profitable urban farms. Some suggestions include developing a care ethic that prioritizes community care and capacity building as well as mutual aid networks. It is a call to arms for people to organize and begin implementing grassroots solutions to local problems, rendering large-scale technologies obsolete.

The methods appendix offers detailed explanations of interview protocol, questions asked, note-taking strategies, personal reflections from the field, and personal anecdotes. In addition, I discuss how I accessed my research sites as well as intersectional aspects of my identity, or positionality, and how they relate to power and privilege in my research.

1
In the Shadow of Sustainable Development

> The environmental crisis is an outward manifestation of a crisis of mind and spirit. There could be no greater misconception of its meaning than to believe it is concerned only with endangered wildlife, human-made ugliness, and pollution. These are part of it, but more important, the crisis is concerned with the kind of creatures we are and what we must become in order to survive.
>
> —Lynton K. Caldwell (as cited in Guimaraes 2004, 460)

The process of modernization is built on the assumption that humanity is on a linear march toward some idealized concept of progress. In this view, progress means moving toward some societal apex through technological advancements and development that grow economies and enhance the lives of human beings by granting more individual freedoms and higher material standards of living (Rostow 1959; Wessels 2013; Mies and Shiva 2014). Modernization proponents at once venerate Western cultures and devalue traditional and Indigenous cultures, presuming the transition from "traditional" to "modern" to be a natural evolution.[1]

Globalization—by which I mean the international movement of corporations, culture, people, and governance—has been integral to modernization since the first European trading companies set up posts near the Indian Ocean in the 1600s. Since the 1950s, Global North states, international monetary agencies, and multinational corporations have worked in partnership with less-developed nations to help them "catch up" by providing loans, securing contracts with multinational corporations, and building infrastructure (Mies and Shiva 2014). This collaboration promotes the rapid development of industrial agriculture and urban areas in societies around the globe. However, Northern elites are driving the globalization agenda and securing disproportionate benefits (Stiglitz 2003). As such, development largely benefits multinational corporations, who take advantage of cheaper

Surviving Collapse. Christina Ergas, Oxford University Press. © Oxford University Press 2021.
DOI: 10.1093/oso/9780197544099.003.0002

foreign labor and raw materials, negotiate business-friendly policies in host countries, and expand consumer markets to poorer nations by exploiting latent demand. Development does allow growing middle-class populations the world over more access to consumer items, including electronics, exotic foods, and textiles. But the promise that everyone could live an affluent and luxurious life is a false one. Indeed, it would be physically impossible for the more than seven billion people on the planet to live like affluent people of the Global North. There's simply not enough space and resources on earth to support that much consumption (Smil 1994; United Nations Environment Programme [UNEP] 2012; Mies and Shiva 2014). Despite some enhancement in quality of life, rapid development has had deleterious consequences for many peoples and our ecosystems globally.

The United Nations World Commission on Environment and Development published *Our Common Future*, also known as the Brundtland Report, to promote environmentally sustainable development through international interdependence and multilateralism (UN 1987). Sustainable development is intended to improve quality of life and advance economies, while conserving natural resources and reducing environmental degradation. Since then, leaders of the Nordic countries, with Swedish policymakers at the helm, have advocated for sustainable development that advances a new form of modernization termed "ecological modernization." Ecological modernization aims to reform capitalism and correct for old, environmentally destructive practices by employing the newest and most efficient technologies (Lidskog and Elander 2012). The UN has since identified seventeen sustainable development goals for member nations. These include ending hunger, reducing economic and gender inequities, taking climate action, building sustainable cities, and establishing industry and innovation, to name a few. Each goal is admirable and important; however, since the 1980s environmental conditions have worsened and social inequities have widened across the globe. If this is what development looks like under ecological modernization—runaway climate change and rampant social inequalities—then modernized development is failing us.

This book discusses the limitations of modern concepts of progress and sustainable development and offers for consideration alternative models that reintegrate agriculture into urban spaces, thereby (re)asserting some autonomy from the current globalized food system. While the alternative models are not prescriptive, they do present a different way to think about the meaning and purpose of progress. The cases I discuss provide examples of adaptive strategies that also have mitigating effects on climate change. In the chapters that follow on the Pacific Northwest ecovillage and Cuban urban farm, I describe approaches to more holistic and encompassing solutions that address social equity and environmental protection. While neither case perfectly manifests egalitarianism or zero environmental exploitation, they each provide an opportunity to learn from and adapt different socioecological technologies.

To explain the paradox of sustainable development, I delve into critical theories of environmental sociology, ecofeminism, and environmental justice. These theories expose the root causes of our socioecological problems and clarify the necessity for holistic solutions that are at once transformative—dismantling hierarchies—as well as regenerative—healing and restoring the health of people and the planet. I argue that creative and radical approaches that get at the root of socioecological problems are needed now more than ever. Indeed, what is needed is a radical sustainability that is at once socioecologically transformative and regenerative, whereby the economic system works in service of social equity and environmental restoration. Globalization has spread socioecological problems direly around the world. The international community has experimented with the reformative ecological modernization approach the since the 1990s, and this approach is not working. Carbon emissions actually increased in 2018, and the Intergovernmental Panel on Climate Control (IPCC 2018) warns that we have just over a mere decade to take action against catastrophic climate change (Le Quéré et al. 2018). It is time to rethink our notions of progress.

The model cases will be explored in depth in the chapters that follow. Here I provide the context for our current socioecological crisis, guided by three questions: How did we get here? What are current debates about potential solutions? What barriers exist to socioecological change? I begin with a brief history of urbanization and industrial agriculture, then define sustainable development, offer insights from critical environmental justice and ecofeminist sustainability scholars, explain the constraining features of capitalism in meeting UN goals, and, finally, discuss the possible roads to recovery.

1.1. How Did We Get Here?

1.1.1. Urbanization and Rural Extraction

As a testament to the transformative powers of our modern global economic system, more than 50 percent of the world's population now lives in urban settings, and the United Nations predicts that this number will continue to grow (UN 2009). Urbanization contributes to many global environmental problems, including climate change, deforestation, food insecurity, and environmental injustice. It necessitates extractive developments in rural hinterlands, for accessing food, energy, water, and other resources, and extending transportation. For example, urbanization requires the rural development of industrial agriculture, because people in cities have little or no space for farming. The mechanization of agriculture greatly increases food production while requiring less human labor, which contributes to population growth and provides urban laborers. Research indicates that industrial agriculture contributes to topsoil erosion, salinization, excess nutrient runoff, dead zones, and climate change (WWF 2016). Resource extraction and transportation also contribute to climate change and air and water pollution, among other problems. In this section, I briefly outline the processes of urbanization that have brought human society to a mostly urban and progressively more treacherous world.

1.1.1.1. Industrialization and the Rise of Modern Cities

Although the ancient precursors to cities date back to 3000 BCE, modern cities are distinctly capitalist constructs (Mumford 1961). Modern and early cities share social hierarchies that prescribe resource distribution and access *and* the exploitation of the natural environment. Preindustrial cities practiced some agriculture within the city and engaged in some trade, but mostly relied on nearby resources. By contrast, modern cities center on trade, and their labor and environmental exploitation are global in reach. Mumford (1961) describes the evolution of cities as "a desire to tame and control nature, to dominate and master strong or mettlesome animals . . . above all, to exercise, partly by command of weapons, a predatory power over other human groups" (Mumford 1961, 21). While some researchers contest his account, many agree that the desire to control nature and certain groups of people remain prominent features of cities today (Merchant 1980; Waring 1999; Davis 2006; McMichael 2010a). Several concurrent socioeconomic changes coalesced to spur the transformation from preindustrial to modern cities, including primitive accumulation, the enclosure of communal land, intensified divisions of labor, and industrialization. While preindustrial cities maintained some internal agriculture, industrialization further severed towns from countryside, spurred rapid population growth, bolstered trade to sustain growing populations, and centralized manufacturing (Ponting 2007; Mosley 2010; Penna 2010).

The Industrial Revolution, which began in England around 1760 CE, made industry and trade paramount to urbanization. Indeed, in the eighteenth and nineteenth centuries many company towns grew around extractive industries and were politically controlled and owned by the local industry (though fewer now, many company towns remain; Green 2010). Mechanization and advancements in energy production, such as coal-fired steam power, rapidly expanded commodity production and transportation for consumption and trade. With increases in food and resource production, urban populations grew. For example, in 1851, 40 percent of Britain's population was urban; by 1900, 75 percent lived in cities. Over the last two centuries, the global urban population has increased 107-fold, with the most growth occurring in the last half century (Ponting 2007). Western European trade networks and colonization spread this style of production through Europe and the United States, and later throughout the world.

England's process of industrialization and widespread urbanization was made possible by a combination of agrarian capitalism and the enclosure of the commons (Williams 1973). The "commons" refers to land and resources "owned, managed, and used by the community" and "embodies social relations based on interdependence and cooperation" (Shiva 2005, 21). Enclosures began as a legal process of closing, or fencing off, what had been communal agricultural land for elites' private ownership. However, unofficial enclosures, a process by which the dominant elite seized and controlled common rural lands without parliamentary approval, had been taking place as early as the thirteenth century through conquest (Williams 1973, 96–97). By the first quarter of the nineteenth century in England, parliament had passed thousands of legislative acts that allowed politically dominant landowners to appropriate "more than six million acres of land," or about a quarter of all cultivated acreage (Williams 1973, 96).

From the fifteenth through nineteenth centuries, European colonization encroached on commons all over the world, displacing peoples in the Americas, Africa, and Asia. Since the inception of the first multinational company, the British East India Company in 1600, capitalism has been a colonial project invested in the exploitation of Indigenous labor and land (East India Company 2018). Shiva (2005) asserts that colonization has never ended, and a neocolonial version of enclosures is continuing today under the rhetoric of globalizing modernization. This consists of the depeasantization of the developing world through internationally funded development projects, trade liberalization, and the spread of industrial agriculture to feed growing urban populations (Shiva 2005). The new dominant elites are financial institutions, such as the World Bank and International Monetary Fund (IMF), multinational corporations, international organizations such as the United Nations, and political leaders of the developed world. Globalization is a global enclosure movement that transforms countries, economies, landscapes, and the way people live, and it is responsible for the global rural-to-urban transition (Shiva 2005; Davis 2006; United Nations 2009; McMichael 2010a, 2010b; Glaeser 2011).

The rise of capitalism, beginning in Europe in the sixteenth century, helped spur the shift from rural to urban living. The establishment of world trade and financial markets marked the origins of capital accumulation, and new social relations and divisions of labor began in newly enclosed rural areas. The precursors to capitalist expansion operated on multiple geographical scales simultaneously (Ponting 2007). For example, on a local scale there was the manorial estate; nationally, the creation of national debts; and globally, the African slave trade. Capitalist development has been characterized as the commercial town's victory over the countryside whereby "agriculture . . . becomes merely a branch of industry and is entirely dominated by capital" (Marx and Engels as cited in Moore 2000, 126). David Harvey (1982) argues that the signal, interrelated characteristics of capitalism are human alienation from the means of production, displacement from the land, and population concentration in urban centers. Therefore, under capitalism, people are driven into urban areas, depriving them of access to the land. In particular, landownership and rent are two mechanisms that "prevent labourers from going back to the land and so escaping from the clutches of capital" (381–82). Capitalist institutions discourage both the dispersal of population beyond urban sprawl and the development of new farming communities. Anderson (1976) also recognizes that there is a limit to the dispersal of the population under capitalism, writing: "Decentralization, outside of urban sprawl, is not profitable" (190). Urbanization is integral to modernization because it allows for a smoother funneling of resources, such as water, energy, and food, to captive, paying urban consumers.

Urban theorist Harvey Molotch (1976) refers to cities as "growth machines." He focuses on the political and economic forces guiding land use in urban settlements. By analyzing social structure, power, and class hierarchies, Molotch identifies economic growth as the primary driving force of political will. In this view, the main goal of each city is to attract as much business as possible to grow the city and the economy. As engines of economic growth, cities are materially dependent on resources from hinterlands; thus, they depend on regional or global trade (Rees and Wackernagel 1996). Leaders are businessmen, lawyers, and realtors who have a vested interest in attaining more resources and capital and may not represent the majority of people who make up the city. Thus, social issues like social justice, environmental protection, and labor issues are treated as "symbolic" and auxiliary.

In addition, Molotch and Logan (1984) identify a tension that exists in cities between exchange value (i.e., money) and use value (i.e., functionality; Jonas and Wilson 1999). Many urban developers see the potential for exchange value in each parcel of land. That is, they buy land based on its potential for commodity production and profit generation. Moreover, local politicians generally establish zoning laws and codes to facilitate that exchange. Conversely, because the city is also a living space, urban residents often see the potential use value of urban lots. Instead of an industrial site, residents may prefer a park or community center, spaces that do not generate profits. Because sites like these run counter to the growth machine logic, they are often difficult to obtain or maintain without constant struggle via social movements. If we could reimagine cities predominantly around use value and quality of life rather than exchange value and growth, we could create socially just and sustainable urban environments. Before turning to reimagining cities, let us consider the social and environmental problems associated with urban spaces.

1.1.1.2. Urban Socioecological Problems

Current social and environmental problems are the legacies of European colonizers' exploitation of certain peoples and environments. The architects of industrial cities saw human society as above nature. Colonialism resulted in the theft and enslavement of Indigenous lands and communities for the benefit of European elites. Modern cities have mastered the exploitative practices that gave rise to their prominence, as they ravenously consume labor and resources from all over the globe at accelerated rates. Modern cities also drastically change local environments, so that they are prone to air, land, noise, and water pollution, as well as overcrowding, traffic congestion, heat islands, and concentrations of greenhouse gas emissions (McNeill 2000; IPCC 2018; Melosi 2010). In addition, and not unrelated, today's cities are loci of mounting social and economic inequalities (Molotch 1976; Davis 2006).

Over time, the shift in human populations from rural commons to privatized urban areas came about for one of two reasons: either the people *had* to leave—push factors—or they *wanted* to go to the city—pull factors. Push factors, such as the mechanization of agriculture, civil war, drought conditions, and competition with large-scale agribusiness—have outpaced urban pull factors, like employment, because developing cities are strapped with debt and economic depression (Davis 2006, 16–17). As a result, urban poverty and slum conditions proliferate. Specifically, slums—or settlements characterized by overcrowding, poor housing, inadequate sanitation, and insecurity of tenure (Davis 2006)—are growing faster than cities and the infrastructure necessary to abate abject poverty. More than one billion people, about a third of the world's urban population, now live in slums, which number more than 200,000 worldwide (United Nations Human Settlements Programme 2003; Davis 2006). Further, about half of all urban dwellers live in poverty even if they do not live in a slum (Davis 2006).

Cities are built on the backs of exploited laborers and are still largely segregated, with the elites living on the most desirable land (Davis 1990; Colten 2005). Large concentrations of people in small areas lead to increased local environmental and health hazards. However, elites are able to buffer themselves from environmental hazards and are more likely to live in areas secure from flooding and far from waste disposal sites and industry. The world's poor are typically the last to receive effective drainage, sewage, and water services, if they receive them at all (Colten 2005; Ponting 2007). While most cities of the Global North, like New York and London, found ways to pipe in clean water, treat sewage, and dispose of garbage by the early twentieth century, many cities in developing nations are experiencing population growth rates that exceed the capacity of infrastructure to manage the corresponding growth in waste. Thus, poor urban populations experience health problems associated with industrial effluence, sewage, and trash (Ponting 2007; Mosley 2010; Penna 2010). The lack of sanitation services in slums accounts for 1.6 million deaths per year worldwide. Inequities exist in developed nations as well. Colten (2005) examined how inequity was built into environmental modifications in New Orleans. Public utilities were first allocated to affluent white neighborhoods and deliberately bypassed African-American neighborhoods (Colten 2005). Before New Orleans built extensive sewage systems, poor laborers were paid to empty the contents of privies and ditches into the local waterways (Colten 2005).

Historically, antipollution legislation has been a challenge to pass, because state policy is often designed to protect developing industries even at the expense of people's health and the environment (Molotch and Logan 1984; Mosley 2010). Air pollution kills an estimated 1.5 million people worldwide annually. Some industrial cities, like Lanzhou in China, burn coal for fuel. Lanzhou residents are exposed to over 100 times the particulate limit set by the World Health Organization (WHO; Davis 2006; Mosley 2010). Currently, urban transportation, commercial buildings, and homes in developed nations consume fossil fuels for energy and are among the highest emitters of greenhouse gases. In the developing world, industry and manufacturing in cities are still the major emitters. China recently exceeded the United States as the highest greenhouse gas emitter. However, *residents* of affluent nations produce far more greenhouse gas emissions each year per capita than do those in developing countries. In 2016, United States residents emitted about sixteen tons of carbon dioxide, while China and India emitted seven and less than two tons per capita, respectively (World Bank 2021).

1.1.1.3. Urbanization as Culprit and Savior

Urban enthusiasts claim that cities are not inevitably waste-sinks overrun by pollution. Paradoxically, they can be sites of potential solutions and change by harnessing the spatial density of people for energy efficiency and social movements (Rees and Wackernagel 1996; Mol 2001; McNeill 2000; Mazmanian and Kraft 2009; Owen 2009; Melosi 2010; Satterthwaite 2009, 2010; Glaeser 2011). Many urban researchers acknowledge just how unnatural the process of creating a city is (Cronen 1991; Colten 2005), but some are convinced that these are current but unnecessary conditions for cities. Controversially, Glaeser posits that urban poverty is a sign of a thriving city that is inviting to people who need work. In addition, Rees and Wackernagel argue that "cities and their inhabitants can play a major role in helping to achieve global sustainability" (1996, 223; see also Satterthwaite 2009, 2010).

Glaeser (2011) claims that cities are centers for the spread of ideas that have social movement potential. Recently revisited by activists, Lefebvre's (1996) work on *The Right to the City* details the ways in which residents have the right to collectively reclaim aloof and commodified urban spaces. This reclamation includes relinquishing the more alienating features of cities, such as capitalism and inequality, and replacing them with cocreated community vibrancy, remaking the city as a junction for community life as well as mobilization and collective action. In this way, Lefebvre emphasizes the transformative potential of dense urban living, allowing for social movement against inequitable and unsustainable relationships.

Rees and Wackernagel's (1996) research demonstrates the immense variation in greenhouse gas emissions from highly urbanized nations, pointing to the potential for high standards of living with less pollution. While recognizing that the affluence of a population and the potential for overconsumption creates barriers to urban sustainability, some urban theorists further argue that certain key features of cities, such as urban density, allow for high standards of living with lower emissions (Satterthwaite 2010). Owen (2009) explains that places like Manhattan are home to some of the most energy-efficient people in the United States because they drive less, live in smaller dwellings, and live closer to amenities. Residents have access to public transportation, convenient walking or biking paths, live near goods and services, and live in energy-efficient high-rise apartments. Specifically, 82 percent of Manhattanites in New York commute to work via public transit. At the time of his research, the average New Yorker emitted 7.1 tons of greenhouse gases, while the average American emitted 24.5 tons (Owen 2009). In addition, Rees and Wackernagel (1996) point out that cities can more readily provide treated and piped water, sewer systems, waste collection, material recycling and remanufacturing, and less demand for occupied land.

Satterthwaite (2010) contends that urbanization does not drive environmental degradation but that economic and political activities pushing development and economic growth do. Even though modern cities have been built with an economic growth dictate, not all are equally destructive (McNeill 2000). Havana, Cuba, though built under the colonial industrial model, demonstrates that urban space can be reorganized for agriculture and reforestation (Koont 2011). Melosi (2010) characterizes cities as natural as human beings themselves and subject to the same limits as all ecological systems. Finally, Glaeser argues that individuals should remain in cities because "if you love nature" you should "stay away from it" (2011, 201). The one thing urban researchers agree on is that cities have dramatically changed things—societies, landscapes, economies, and ecologies—for better or worse.

1.2. What Are the Current Debates on Solutions?

Divergent opinions exist on how to move toward socioecological sustainability. Some people believe we can continue to grow the global economy by developing green technologies that allow us to adapt to and mitigate the harmful effects of climate change, such as renewable energies, carbon capture and storage, or even moving some people to Mars. Conveniently, focusing on technological development requires little individual behavioral or social structural change. In contrast, there are those who believe we must halt our population growth and again live like gatherers and hunters, completely minimizing our environmental footprints. Notably, neither of these "solutions" take into account the current nor future social costs of these changes. The likely reality is that changes will need to be made on all fronts, including individual behavioral, legislative, and technological change. Fortunately, we have all the information and tools we need to begin making these changes. Yet, contentious international debates between activists, corporate executives, politicians, and scholars on appropriate solutions toward socioecological sustainability remain.

1.2.1. Sustainable Development

The term sustainable development is vague, heavily debated, and rife with contradiction (Portney 2003; Larsen 2009; Mazmanian and Kraft 2009). Sustainability itself has as many as seventy different definitions (Gaard 2017). The UN Brundtland Commission (1987) put forth the most commonly cited definition of sustainable development, which I define in the introduction. Sustainability, as conceived by the UN Brundtland Commission and *Agenda 21*, is comprised of three overlapping and mutually reinforcing goals: (1) to live in a way that has long-term environmental viability; (2) to live in a way that maintains economic living standards now and into the future; (3) and to live socially just and equitable lives into the future (UN 1987; UN 1992; Moore 2007; Dillard, Dujon, and King 2009). These goals are also known as the "triple bottom line" of equity, environment, and economy, also termed "people, planet, and profit."

While these three goals can be mutually reinforcing, it is often the case that one goal is prioritized at the expense of one or both of the others. For example, there are cases where environmental conservation efforts negatively affect Indigenous communities by displacing them and denying them access to ancestral lands (Neumann 1998; Carruyo 2008) and other cases where toxic polluting industrial practices are overlooked or deregulated despite the known negative consequences on human and environmental health (Carson 1962; Lubitow and Allen 2013). In addition, development projects in poor nations often uproot Indigenous subsistence agriculturalists and pastoralists without adequate compensation, relocation strategies, or alternative employment opportunities (Shiva 2005; Davis 2006; Pellow 2007; Braun 2008). In cases where individuals are promised compensation, it is not uncommon for corruption among local authorities, in partnership with international development agencies, to interfere with the affected parties' ability to receive full compensation (Davis 2006). These examples demonstrate the fraught balancing act between the social and economic aspects of sustainability and the ecological.

The concept of sustainability has its roots in biology and ecology, particularly the notion of ecological carrying capacity. "Carrying capacity focuses on the idea that the earth's resources and environment have a finite ability to sustain or carry life, particularly animal life. Similarly, a particular ecosystem has a finite ability to sustain the life contained there. When the demands move beyond the carrying capacity of the earth or a particular ecosystem . . . species collapse will occur" (Portney 2003, 5). Therefore, for human societies to survive and thrive, human activities, both individual and collective, must account for carrying capacity.

One method that measures carrying capacity is the ecological footprint. The ecological footprint "is the total area of productive land and water required continuously to produce all the resources consumed and to assimilate all the waste produced, by a defined population, wherever on Earth that land is located" (Rees and Wackernagel 1996, 228–29). The footprint estimates the area of land and ocean required to assimilate the waste of a given population and support their consumption of food, goods, services, housing, transportation, and energy (Rees and Wackernagel 1996; Center for Sustainable Economy 2019). Biocapacity refers to the earth's available ecologically productive land, measured in global hectares (gha; Wackernagel, Monfreda, Moran, Wermer, Goldfinger, Deumling, and Murray 2005). When a given population exceeds biocapacity, then they are running on an ecological deficit. This deficit means that either they are importing resources or that productive ecosystems are in decline and the situation is not sustainable over time. Since the 1970s human society has been running on an ecological deficit, where forests, cropland, pastureland, and marine fisheries are being depleted and carbon emissions are higher than can be reabsorbed by the environment (Wackernagel et al. 2005; WWF 2012). According to the *Living Planet Report* (WWF 2016), there are 12.2 billion global hectares of bio-productive land. In 2012, humanity's footprint was 20.1 billion gha, an average of 2.8 gha per person, which is 50 percent more than the earth's biophysical capacity. To live within limits, an equitably distributed share would be 1.8 gha per person. To be sure, the 20.1 billion global hectares of productive land that humanity is using are not equitably distributed. People in the United States average over 7 gha per person, while people in Cuba average just under 2 gha per person (WWF 2012, 2016). The disparity in productive land used per capita is significant, with the biggest consumer, Qatar, at close to 12 gha and several nations, like Madagascar, under 1 gha per person (WWF 2012). Of course, there are also disparities of consumption within nations. However, affluent nations on average consume the most resources and produce the most waste (WWF 2012).

Dietz, Rosa, and York (2007) found that the two main drivers of anthropogenic environmental problems are population size and affluence, controlling for a number of other factors, including urbanization. The IPAT formula specifies that environmental impacts (I) are the product of three driving forces: size of "population [P], affluence [A] (per capita consumption or production) and technology [T] (impact per unit of consumption or production)" (York, Rosa, Dietz 2003a). Their cross-national data analysis reveals that nations with higher gross domestic product (GDP) per capita also have higher ecological footprints. The significance of affluence demonstrates that the impacts of population on the environment cannot be fully understood without considering the scale and mode of production and consumption particular to different societies. These findings and ecological footprint calculations also have implications for the economic and social aspects of sustainability in relation to the environment. I discuss these implications in more depth below.

At the conception of sustainable development in 1987, The Brundtland Commission accepted economic growth as the path to improved human well-being while neglecting to include an analysis of global power relations. In the wake of criticisms, in 1990 the UN developed the Human Development Index (HDI)—"a composite statistic of life expectancy, education, and income indices used to rank countries into four tiers of human development"—to measure human progress (UN 2013). Their purpose was to ascertain the quality of development and improved social well-being, in addition to the standard practice of measuring the quantity of economic growth, like gross national product (GNP). However, critics have pointed out that the HDI, while an improvement, misses a number of important variables that would provide more meaningful and comprehensive information about a nation's overall well-being.

The UN's concerns notwithstanding, our global ecological footprint keeps expanding beyond the earth's capacity to sustain human society, now and into the future, and inequitable wealth distribution continues to grow (Dietz, Rosa, and York 2007; Magis and Shinn 2009). Indeed, Milanovic (2005) demonstrates that international inequality began a sharp ascendance in 1982 and continues to grow into this century, where poor countries do worse on average than rich ones. One can see these trends in several reports. In 1999, the UN observed that income inequality had increased during the post-World War II era; "The income gap between the fifth of the world's people living in the richest countries and the fifth in the poorest was 74 to 1 in 1997, up from 60 to 1 in 1990 and 30 to 1 in 1960" (3). While a 2013 UN report indicates that income inequality is still on the rise, HDI shows progress in health and education access (Yates 2012). Disparities within nations are rising as well. Specifically, researchers in 2011 revealed that "wealth inequality in the United States is at historic highs, with some estimates suggesting that the top 1% of Americans hold nearly 50% of the wealth, topping even the levels seen just before the Great Depression in the 1920s" (Norton and Ariely 2011). These trends continue to worsen. In a recent report released by The Institute for Policy Studies (Collins, Chuck, and Hoxie 2017), researchers documented that "the three wealthiest people in the United States—Bill Gates, Jeff Bezos, and Warren Buffett—now own more wealth than the entire bottom half of the American population combined" and "America's top 25 billionaires . . . have as much wealth as 56 percent of the population" (2).

Social inequalities affect democracy and resource access, as inequality commonly expresses itself as a hierarchical relationship wherein resources such as environmental goods, power, and wealth are unevenly distributed. Environmental risks and vulnerabilities are also unevenly distributed, disproportionately burdening those with less power and status. As Pellow (2014) observes, "inequality is a form of domination and control over people, nonhumans, ecosystems, the planet, and life itself. Rising inequality means that most of us are losing control over our ability to influence our own destinies and protect the people and nonhuman relations we care about" (10).

Consequently, sustainable development has fallen short on some important goals.

1.2.1.1. Radical versus Strong versus Weak Sustainability

Sustainable development, particularly for developing countries, is supposed to be the answer to problems of poverty and the environment (Portney 2003, 7). However, as I noted, the global community, particularly in the developed world, has been running on an ecological deficit since the 1970s, and the trends are worsening (Wackernagel et al. 2005; WWF 2012). In addition, markets are unequal in ensuring access to resources, and development since the mid-twentieth century is largely responsible for the massive urban migration patterns that exacerbate urban poverty in less developed countries today (McMichael 2010a, 2010b).

Since the 1992 United Nations Conference on Environment and Development (UNCED) published the "Rio Declaration on Environment and Development" as a guide, sustainable urban development has become institutionalized as part of general urban planning. As a result, international development agencies have partnered with multinational corporations to mainstream the rhetoric of sustainable development, and they purport to consider equity and the environment in their projects. These international organizations work together to finance development that attempts to integrate Global South populations into the global economy by establishing specialized economic niches and global market dependency. This, in part, entails urbanizing certain regions for commerce, building the supporting infrastructure, converting rural lands into large-scale farms or resource extraction zones for raw material export, and, ultimately, displacing peasant farmers.

While the UNCED initially inspired optimism and direction for many environmentally conscious planners and policy makers, sustainable development is a business enterprise (Brand and Thomas 2005). Corporate interests and economic elites have hijacked the rhetoric of sustainable development, while continuing to prioritize economic gain at the expense of ecological and social justice concerns. To ecological theorists (Catton and Dunlap 1978; Commoner 1992; Foster 2000) the interests and goals of capital, namely accumulation, growth, and expansion, are antithetical to environmental sustainability, given that natural resources are finite (Brand and Thomas 2005). In this view, sustainable development as it has been conceived is actually a shell game for creating neocolonial dependency in the developing world rather than a great way for multinational corporations to help bring about more sustainable self-sufficient nations. However, ecological modernization, the academic tradition that theorizes and empirically investigates the capitalist form of sustainable development, suggests that technological innovation driven by competition can overcome the limitations of the environment (Mol 2001; Friedman 2008). Indeed, these theorists argue that modernization and development are necessary for ecological sustainability.

Proponents of reforming capitalism ignore deeply rooted structural problems endemic to capitalism. To reformists, major avenues for advancement include developing renewable energies, increasing efficient technologies, implementing market-based carbon cap-and-trade policies, and curbing individual consumption (Friedman 2008). However, environmentalism under a capitalist economic paradigm ignores nature's intrinsic value, instead reducing it to market logic and exchange value. Beyond privatizing land for property, transnational corporations privatize and commodify resources that sustain humanity's basic needs. Examples include water, through sewerage services and hydroelectric stations, air pollution and carbon trading; biodiversity, by patenting genetic resources; and ecosystems, through ecotourism and conservation efforts (Brand and Thomas 2005, 13). Privatizing resources allows for unequal distribution and access because those who can afford more will pay for more resources.

Critics point to the "Netherland's fallacy" and capitalism's logic of growth as the main limits to modernization theories of environmental sustainability[2] (Molotch 1976; Gould, Pellow, and Schnaiberg 2008; York, Ergas, Rosa, and Dietz 2011). Most environmental improvements within nations as they modernize typically occur because nations have shifted the environmental impacts beyond their borders. Rather than engaging in genuine environmental reforms, nations in the process of modernizing import natural resources and export pollution. This assumption that environmental degradation is geographically confined within the boundaries of a nation and that nations are responsible for their own extraction, production, consumption, and waste disposal is called the "Netherlands fallacy" (Ehrlich and Holdren 1971). This refers to the observation that while the Netherlands is known for its decline in environmental impacts at home, it imports most of its natural resources, so the environmental impacts it causes occur in other (typically developing) nations. The Netherlands fallacy exemplifies a process common in affluent nations.[3]

Rees and Wackernagel (1996) propose that for urban development to be sustainable, advocates must ensure a "strong sustainability" that stands in contrast to "weak sustainability." "Weak sustainability" operates on the assumption that "manufactured capital can substitute for natural capital" (Rees and Wackernagel 1996, 226). Weak sustainability programs or policies tend to stress economic gain while neglecting either or both equity and ecology, two of the three crucial elements of sustainability. Corporations regularly propagate weak sustainability solutions, and intergovernmental development organizations often partner with multinational corporations to advance these projects. Corporations prioritize profits even when social equity and ecological restoration are among their stated goals (Stiglitz 2003; Davis 2006; Foster 2008; Parr 2009; Rogers 2010). This happens because they are serving the interests of their investors, who expect short-term returns. Investors are often disconnected from, and their interests are in conflict with, the populations the business serves, the poor, and the accompanying natural environment. The result is "green-washed" consumerism—unsustainable business practices disguised as ecofriendly.

Greenwashing is common practice under weak sustainability and includes green marketing, nonbinding environmental agreements, and market-based regulation of pollutants, rather than taking real steps toward environmental change. As an example, a large company in Paraguay called Azucarera Paraguaya provides two-thirds of all organic sugar consumed in the United States and distributes to big names like Imperial Sugar and Whole Foods. Azucarera employs monocropping methods, which scientists argue diminish soil nutrients, cause soil erosion, and deplete groundwater. They also rely on the deforestation of one of the most threatened forests in the world—the Upper Paraná Atlantic Forest—where only 8 percent of the ecosystem remains. In addition, this company uses the manure of industrially raised chickens, which are given antibiotics and feed laced with arsenic to promote growth. Such environmentally destructive practices remain common because organic seal-certifying bodies are often hired by the companies they regulate, and hence may require little accountability (Rogers 2010). Racial and economic inequalities also are often perpetuated or exacerbated by sustainable development projects. The case of recycling in Chicago is an example of this type of weak sustainability. When Chicago introduced recycling centers in low-income communities of color, the centers helped reduce the amount of waste going into landfills while creating jobs, but exposed employees from the local community to dangerous and toxic working conditions (Pellow 2002).

Rees and Wackernagel argue that cities should strive for "strong sustainability" by operating under ecological principles that include adhering to biophysical limits and recognizing that "biophysical capital perform[s] critical functions that cannot be replaced by technology" (1996, 226). In terms of ecology, Sadler (1999) posits that strong sustainability entails "maintaining natural capital at current levels (no net loss). The resource losses and ecological damages resulting from development must be replaced or offset." Finally, he maintains that "basic human needs must be met everywhere" (21–22). Strong sustainability differs from weak sustainability in terms of emphasis, specifically stressing the equity and ecology aspects of sustainability as an indicator of economic gain. Although the enduring health of an ecosystem is never guaranteed, strong sustainability includes prioritizing the needs of the most vulnerable populations of people while maintaining the natural environment (Rees and Wackernagel 1996). Policy for sustainable development should include sustainable use of hinterlands, which means ensuring that "natural capital stocks are adequate to provide the resources consumed and assimilate the wastes produced by the anticipated human population into the next century, while simultaneously maintaining the general life support functions of the ecosphere" (Rees and Wackernagel 1996, 226). Sustaining rural hinterlands should be done while ensuring that resources are distributed evenly and in a socially just manner.

Most modern cities are not sustainable on their own, but some researchers argue that cities can be maintained if urban areas do not exceed the ecological limits of their corresponding hinterlands (Rees and Wackernagel 1996). To date, there are few examples of cities living within ecological limits in nations with high human development,[4] and there are no contemporary examples of such cities with high material standards of Western living (Satterthwaite 2010). Most current examples of cities with high material standards of living operate under the capitalist, pro-growth model of "weak" sustainable development (Rees and Wackernagel 1996). Since there are no examples of cities with high material standards of living operating under a "strong sustainability," it remains to be seen how sustainable any city can actually be.[5]

While the strong sustainability that Rees and Wackernagel put forward is important, I argue that their articulation of strong sustainability still falls short because it does not call into question power imbalances and the root cause of environmental degradation and social inequity, the global economic system. In addition, strong sustainability doesn't push us to give back to the ecosystems that sustain us so that they may flourish beyond meeting humans' needs. I suggest that we attempt what environmental activists have long fought for, and what critical environmental justice and ecofeminist scholars have written about: a radical sustainability that includes transformative—dismantling hierarchies toward total liberation—and regenerative—healing and restoring the health of people and the planet—socioecological relations (Kellogg and Pettigrew 2008; Pellow 2014; Gaard 2017). We should be trying to leave the planet a better place than when we found it!

Radical sustainability challenges us to resolve the root causes of socioecological crises—the global growth economy that relies on the exploitation of peoples' labor and the environment—by fostering equity and regenerative ecological practices. Radical sustainability is community oriented (yet celebrates individual and cultural difference), encouraging self-sufficient communities that are premised on ecological principles. Radically sustainable economics emerges from socioecological interdependence and reciprocity. It may involve figuring out the most effective and socially just use of local land and resources through directly democratic decision-making processes. Concurrently, communities must make efforts to support interrelated ecosystems and minimize pollution by maintaining closed-loop systems, like nutrient cycles. Environmental stewardship should be prioritized among human activities. Economic gain and progress can be (re)envisioned to include increases in quality-of-life indicators and the health of the ecosystem, similar to the Happy Planet Index (2016). Relocalization of most environmental resource use is necessary, because it promotes local community participation in every step of the production, consumption, and waste disposal process. This keeps the community self-sufficient and aware of the effects of their activities because they can see, and adapt to, changes in their local environments. Capitalism as a global system of inequality is not likely to deliver a radical sustainability. Focusing on the social aspects of sustainability is important for a radical sustainability because a growing body of research has demonstrated that social inequalities can lead to environmental degradation, which in turn harms vulnerable human communities (Ergas and York 2012; Pellow 2014; Cushing et al. 2015).

1.2.1.2. Social Sustainability

The UNCED in Rio de Janeiro is widely regarded as the global community's acceptance, and mainstreaming, of sustainable development. This conference published *Agenda 21* and the "Rio Declaration on Environment and Development" as guides to sustainable development. The first principle of the Rio Declaration (1992) states, "human beings are at the centre of concerns for sustainable development." The social aspect of sustainability is the least theoretically developed and empirically studied of the triad, and therefore is weakly articulated (Larsen 2009). However, a cultivated tradition of research on social well-being exists (Magis and Shinn 2009). It includes both intra- and intergenerational equity—improving the well-being of all people worldwide, especially the poor, so that they *thrive* now and for generations into the future (Sadler 1999; Magis and Shinn 2009). If development is human centered, then indicators of human well-being should be central to all indicators of progress. However, growth in GDP is still the most valued indicator of development (Waring 1999; Milanovic 2005; Magis and Shinn 2009). Here, I review the literature on social sustainability.

According to Jonathan Harris and Neva Goodwin, "A socially sustainable system must achieve fairness in distribution and opportunity, adequate provision of social services, including health and education, gender equity, and political accountability and participation" (Harris and Goodwin 2001 as cited in Dillard, Dujon, and King 2009, 3). Dillard, Dujon, and King add two aspects: "the processes that generate social health and well-being now and in the future" and "those social institutions that facilitate environmental and economic sustainability now and for the future" (2009, 4). These definitions move beyond society's direct relationship to the environment, like property rights, environmental knowledge and ethics, and land and resource tenure, to a discussion of how society can be sustained in its own right. Magis and Shinn (2009) discuss the four interrelated principles that emerge out of international discussions of social sustainability—these principles are human equity, well-being, democratic government, and democratic civil society. I examine each of these principles.

1.2.1.3. Equity

Equity encompasses multiple dimensions, including gender, income, and inter- and intragenerational components. Specifically, equity means accessible economic and political opportunities for all people through redistributing power and wealth and eliminating inter- and intragenerational disparities within and between nations (Magis and Shinn 2009, 22). Gender equity also is an important aspect of this principle. *Agenda 21* (UN 1992) includes chapter 24, entitled the "Global Action for Women Towards

Sustainable and Equitable Development." It states that the effective implementation of *Agenda 21* requires that women must be actively involved in economic and political decision-making (UN 1992). Research has shown that women's significant presence in community decision-making bodies leads to better protection of available environmental resources. My own cross national research demonstrates that nations with higher women's political empowerment tend to have lower CO_2 emissions (Ergas and York 2012). The UN suggests some strategies to promote women's political participation including ensuring women's access to education relevant to their roles in resource management, rights over their reproductive choices, safe and affordable healthcare, and consumer awareness programs (UN 1992).

Other forms of social inequality aside from gender also potentiate the social divisions that are linked to poor environmental quality. For instance, racial residential segregation lengthens commutes and vehicle miles traveled, which worsen air and water quality overall (Morello-Frosch and Jesdale 2006; Morello-Frosch and Lopez 2006). In addition, income inequality is correlated with greater environmental harm (Pellow 2014). For instance, increased income inequality is associated with greater biodiversity loss, resource consumption, waste generation, toxic emissions, water pollution, and lower survival rates of children under five (Cushing et al. 2015; Islam 2015). Inequality not only leads to disparities in environmental exposures that disproportionately burden the disadvantaged, but inequality also leads to higher overall levels of exposure to health-damaging pollutants for everyone (Mohai, Pellow, and Roberts 2009). Therefore, eradicating myriad social inequalities plays a critical role in advancing and realizing sustainability goals.

Magis and Shinn (2009) charge inequality with environmental damage and argue that "sustainability absolutely requires a concerted focus on eradication of inequities" (33). IPAT and ecological footprint analyses show the far-reaching environmental consequences of affluence, but scholars argue that poverty has a unique effect on the environment as well (UN 1987; UN 1992; Moore 2007; Dillard, Dujon, and King 2009). The UN (1987, 55) highlights the fact that poverty negatively affects local environments by positioning poor people "to overuse environmental resources" for day-to-day survival. "Poverty itself pollutes the environment, creating environmental stress in a different way. Those who are poor and hungry will often destroy their immediate environment in order to survive: They will cut down forests, their livestock will overgraze grasslands; they will overuse marginal land; and in growing numbers they will crowd into congested cities" (55). While research shows that indicators like increasing affluence, or GDP per capita, are correlated with environmental degradation, other human development indicators like access to education and increased life expectancy show no relationship to environmental harm (Dietz, Rosa, and York 2007). These findings suggest that development oriented at increasing quality of life rather than GNP or GDP will better serve the stated goals of sustainable development.

1.2.1.4. Well-Being

Magis and Shinn define human well-being as "the fulfillment of basic needs and the exercise of political, economic, and social freedoms" (2009, 16). Basic needs include those important for physiological function, food, water, shelter, and sanitation. These are basic human rights and necessary because humans can't function without them (Magis and Shinn 2009). Others add that people should be able to achieve self-actualization, or the realization of their full potential (Stricker 2007). To this I add Alexander's (2008) insights on psychosocial integration, which include the interdependent human needs of individual freedom and social belonging.

Most theorists focus on the individual aspects of social well-being; however, to have a thriving civil society, we must also account for community well-being. According to Alexander (2008), psychosocial integration "reconciles people's vital needs for social belonging with their equally vital needs for individual autonomy and achievement" (58). It is experienced at once as identity, meaningful community involvement, oneness with nature, and connection with the divine. Alexander suggests that psychosocial integration can be used interchangeably with terms such as social cohesion and community. He also argues that "an enduring lack of psychosocial integration, which is called 'dislocation' . . . , is both individually painful and socially destructive" (58). Individuals need to find and establish a place in their communities, and thriving communities have dense social networks, collaborative work toward community betterment, and social trust and cohesion, also known as social capital. The relationship between the individual and community is synergistic and contains feedbacks. If a community lacks economic opportunities, adequate education, or social spaces, individuals are more likely to express shame, depression, anxiety, addiction, and suicide. As individuals turn inward, they are more likely to lash out violently or act selfishly, which further disintegrates social trust and cohesion. Thus, maintaining psychosocial integration and community cohesion as well as individual well-being and equity are important for a thriving civil society. Indeed, these are closely related to democratic participation because populations need adequate provisioning and access to political processes in order to participate. As an example, Polayni (2001) reveals how discarding systems of social welfare in conjunction with liberalizing the market in Britain undermined societal well-being and led to a declining civil society.

1.2.1.5. Democratic Governance and Civil Society

Magis and Shinn (2009) posit that democracy requires human rights and "access to information, full inclusion, participation, and collaboration" (34–35). A democratic government is one that "ensures that governance is oriented to the people," and where government institutions are "open, transparent, accountable, and supportive of community action" (34–35). They argue that government should serve a protective function, ensuring basic needs, rights, freedoms, public goods, and regulating against economic externalities and harmful market fluctuations. Correspondingly, civil society serves the primary purpose of "ensur[ing] that government is functioning according to the will of its people" (34–35). Civil society accomplishes this by creating civic space and empowering people to use the space for democratic practice, including interaction, coalition building, collaboration, activism, and volunteerism.

Insights from research on social capital posits that social capital is necessary for a functioning civil society (Scott et al. 2012).[6] Putnam (1995) defines social capital as "the features of social organization, such as networks, norms, and social trust that facilitate coordination and cooperation for mutual benefit" (67). Social capital is the mortar of a community and is necessary for social cohesion and reciprocity. High social capital is often found in rural communities that are homogenous, have strong religious affiliations, and dense kinship ties. Conversely, unemployment, depopulation, poverty, dependence on limited economic sectors, low educational attainment, and immense social inequity inspire social distrust and threaten to disrupt social networks and community bonds. Erosion of social capital makes it difficult for communities "to organize to solve collective problems or resist domination" (Scott et al. 2012, 404). Then, disintegration of social capital is a sign of social unsustainability. As the community loses social cohesion, it becomes less resilient and less capable of adapting to disasters, whether the source is social, economic, political, or environmental. Another way to think about social cohesion and social capital is through the concept of psychosocial integration. When individuals lack a means of meaningful community participation, they experience dislocation, which can lead to addiction crises and higher rates of suicide (Alexander 2008). Dislocation is an aspect of eroding social capital and is bad for civil society.

Nascent research exists that examines the links between democracy and sustainability. Early research shows that democratic participation is correlated with certain positive environmental outcomes (Norgaard and York 2005; Winslow 2005; Boyce 2008; Ergas and York 2012). Specifically, Islam (2015) found that efforts to instill sustainability are enhanced by women's participation in community decision-making bodies, leading to better protection of common property resources. Norgaard and York (2005) looked at environmental treaty ratification cross-nationally and the percentage of women holding seats in parliament. They found that nations with higher proportions of women in parliament ratify a greater number of environmental treaties. Similarly, Ergas and York (2012) found that nations with higher women's representation in parliament tended to have lower CO_2 emissions, controlling for urbanization, GDP per capita, and other factors. Another study corroborates these findings, stating that "the higher the level of democracy, the lower the ambient pollution level" (Winslow 2005 as cited in Moore 2007, 188–89). Moore argues that democracy and social equity are good for the environment because "not only do multiple perspectives see the landscape more thoroughly, but they challenge each other to be more innovative" (Moore 2007, 188).

However, there have also been authoritarian and top-down environmental initiatives with beneficial outcomes (Moore 2007; Dillard, Dujon, and King 2009; Mark 2013). Some research demonstrates that democratic governance is not a necessary precondition to sustainable development, as the case of Curitiba, Brazil illustrates. Under a military dictatorship regime from 1964 to 1985, mayor Jaime Lerner of Curitiba enforced a technocratic and authoritarian approach to urban sustainability (Moore 2007). During Lerner's incumbency, Curitiba built fifty square meters of green space per citizen, created a highly regulated and thriving industrial district, and implemented a high-speed bus transit system used by 28 percent of car owners daily, which lowered pollution and respiratory problems compared to elsewhere in Brazil (Moore 2007). Others have highlighted the benefits of top-down regulatory approaches to environmental legislation, particularly the US banning of DDT during the 1970s, perhaps as a response to the demands of civil society (Mark 2013).

Despite prominent counterexamples, Moore (2007, 108–109) argues that participatory democracy is better equipped to maintain urban sustainability for a multitude of reasons. Specifically, elites often insulate themselves from environmental burdens and hazards. Problems that require collective action are more efficiently solved by democratic horizontal networks rather than autocratic vertical networks. Elites tend to suppress conflict to maintain social order; however, conflict is necessary because it provides the feedback loops that identify problems and solutions. Citizens' plans better reflect their needs. If citizens plan it, they will insist on implementation, and they will push officials to stay the course. Finally, Moore contends, "environmental degradation is simply an indication that the interests of some stakeholders are being ignored" (2007, 110).

Indeed, Gaard (2017, 16) argues that democracy is an important feedback mechanism between humans and the environment, and it provides a corrective for ecological problems. She expands on Plumwood's (2002) discussion of five types of remoteness—spatial, consequential, communicative, temporal, and technological—and how they disrupt this feedback and the connections between decisions and consequences. Similar to Moore (2007), she notes that spatial and consequential remoteness—decision-makers living far away from sacrificial ecological zones who do not bear the brunt of negative externalities—allow them to justify the environmental harm because they may benefit from it and do not suffer, or witness, the consequences. For example, corporate and state elites who experience spatial, consequential, temporal, and technological remoteness may financially benefit from the Dakota Access Pipeline technologies that transfer oil across the United States. However, the Sioux Tribe in the North Dakota Standing Rock reservation, who survive on the water the pipeline threatens to contaminate, bear the burden of leaks and spillages in their communities. The Sioux Tribe, in this case, experiences a communicative or an epistemic remoteness, where they have little or blocked communication with elites, thus limiting the transfer of knowledge and motivation for socioecological repair.

Critics of sustainability studies allege that they suffer from anthropocentrism and an undertheorized conception of socioecological justice. Because sustainability is used so disparately by so many different disciplines, it also lacks a cohesive meaning and course of action (Tierney 2015). Additionally, the work tends to be reformist rather than transformative. In order to bridge these gaps, I review just ecofeminist, critical environmental justice, and total liberation frameworks (Pellow 2014; Gaard 2017).

1.2.2. Toward Socioecological Sustainability and Justice

As noted above, social sustainability is the least theorized among the three components of sustainable development. However, critical environmental justice and ecofeminist scholars have insights to nourish this underfed concept. Foster, Clark, and York (2010) identify the root cause of global socioecological problems as "at bottom, the product of a social rift: the domination of human being by human being" (47). Pellow (2014) elaborates on what inequality entails:

> [I]nequality means that if you are "on top" of, or higher on the social ladder than someone else, then you possess or have access to greater resources, wealth, and social privileges. But more importantly—and from the standpoint of ecological politics—your elevation above others also means that your life is of greater value than others living within that social system. You likely own or control and affect more of the planet and its constituent ecosystems than others, you likely own or control and affect more living beings (and therefore likely produce more death) than others, and you likely control and benefit from the ideational systems that give meaning and legitimacy to such dynamics. Inequality is a means of ordering the human and nonhuman worlds for the relative benefit of some and to the detriment of others. . . . Life expectancy, morbidity, mortality, and well-being are highly correlated with key measures of human inequality. . . Inequality is, above all, unnatural in the sense that it does not "just happen"—it requires a great deal of energy, labor, and institutional effort to produce and maintain unequal societies. This point is crucial because there is also so much energy invested into making inequalities seem like a natural state of affairs. . . . Inequality is not just an imbalance of resources or power but is frequently experienced as unearned privileges made possible by domination and injustice. It is also routinely resisted by those who suffer its consequences (6–7).

This detailed definition of inequality contributes crucial components toward a robust theory of social sustainability. Inequality is not just the uneven distribution of resources and harms, but is based on value systems that legitimize unequal access. Inequality must be maintained by "ideational systems," such as the media, as well as through great institutional energy and labor, such as military or police enforcement. The last point, that inequality is "resisted by those who suffer its consequences," is essential for understanding why social inequity is unsustainable. It conjures images of activists chanting in the streets, demanding that we "respect existence or expect resistance." Pellow alludes to the fact that inequity is likely to cause social contention or upheaval, which runs counter to sustaining or maintaining anything. Social upheaval that escalates and results in violence or civil war has grave social and environmental consequences. However, as Gaard (2017) reminds us, if we focus on meeting the needs of the most exploited and marginalized among us—plants, animals, caretakers, peasants, and Indigenous communities—then we actually serve to lift everyone else up along with them in the process. Indeed, research has consistently shown strong relationships between greater social equity, democracy, and better environmental and health outcomes (Agyeman, Bullard, and Evans 2003; Ergas and York 2012; Pellow 2014; Downey 2015). These definitions acknowledge the intrinsic value of the nonhuman world and see our justice struggles as interconnected.

These scholars propose that all oppression is linked, whether related to plant, nonhuman animal, or human. Pellow (2014) explores these linkages to develop the concept of what he calls "total liberation," which "stems from a determination to understand and combat all forms of inequality and oppression" (5). He puts forward a total liberation framework that he argues "animates earth and animal liberation movements" and is bolstered by four pillars: "(1) an ethic of justice and anti-oppression inclusive of humans, nonhuman animals, and ecosystems; (2) anarchism; (3) anti-capitalism; and (4) an embrace of direct action tactics" (5–6). He identifies these linked axes of oppression as socioecological inequality, which he defines as "the ways in which humans, nonhumans, and ecosystems intersect to produce hierarchies—privileges and disadvantages—within and across species and space that ultimately place each at great risk" (7).

With this framework, Pellow attempts to move beyond the human/environment and culture/nature binaries to reveal the importance of intersectionality—or "the idea that oppression cannot be reduced to one fundamental type of inequality" based on racism, classism, heterosexism, colonialism, speciesism, and so forth (Crenshaw 1991; Collins 2000; Pellow 2014, 5). Rather, all forms of oppression are interlocking, mutually constitutive, and reinforcing. He identifies the three primary vectors of socioecological inequality as: "humans exploiting ecosystems," "humans exploiting other humans," and "humans exploiting nonhuman animals" (8). Harm is multidirectional, in that the groups causing harm, the target of said harm, and innocent bystanders may all experience direct or indirect, or feedback, harm. As an example, oil company executives who make decisions to continue oil exploration may also experience heightened wildfire threats to their homes or properties as a result of climate change. Indeed, many innocent humans and nonhuman creatures will disproportionately endure the adverse effects of these executives' decisions. However, because of power and privilege, these same oil executives will suffer disproportionately less loss than poorer individuals who lose their homes or nonhuman animals who lose their habitats. Informed by this framework and the work of ecofeminist, environmental justice, political economist, and critical animal studies scholars, Pellow (2018) proposes what he terms a critical environmental justice.

1.2.2.1. Critical Environmental Justice.

Critical environmental justice builds upon environmental justice scholarship to address its limitations. Environmental (in)justice (EJ) scholarship has largely investigated how environmental goods and harms are unequally distributed between groups of people based on race, class, gender, and nation. Critical environmental justice proposes attention to four more pillars that bolster the work of EJ scholars and activists (Pellow 2018, 17–18). These are:

1. Understanding that "multiple social categories of difference are entangled in the production of environment injustice, from race, gender, sexuality, ability, and class to species," and that state-corporate violence has intersectional effects
2. Embracing multi-scalar, both spatial and temporal scales, methodological and theoretical approaches to EJ struggles;
3. Understanding that socioecological inequity is deeply entrenched in society and therefore, a transformative anti-authoritarian approach to state and corporate power is needed toward environmental justice; and
4. Focusing on the ways in which human and more-than-human actors are indispensable for building the present and future just and resilient communities.

The critical environmental justice framework also has built upon ecofeminist theories integral to a socioecological justice framework.

1.2.2.2. Just Ecofeminist Sustainability

Gaard (2017) borrows from Agyeman, Bullard, and Evans (2003) the term "just sustainabilities"—that emphasize sustainability, environmental justice, and equity—to describe a just ecofeminist sustainability. She advances a conversation between sustainability, environmental justice, ecofeminist, postcolonial, Indigenous, and critical plant and animal scholars. She posits "a sustainable relationship to the environment is linked to care and justice for other animals, women, people of color, queers, and other 'others'" ("The Sydney Declaration on Interspecies Sustainability" as cited in Gaard 2017, 17). In addition, she notes that a "socially responsible sustainability begins where animal exploitation ends" (17). Gaard argues that feminist methodology has always asked questions at micro and macro analytic scales, rooted in a relational standpoint, to observe linkages between personal and political inequalities. She suggests we promote "just and equitable relations by raising questions such as, who benefits, and who pays?" (19).

Critical ecofeminism sees the intrinsic and ecological values of nature, humans as part of ecology, all forms of labor in an economy, and attempts to abolish social hierarchies. In other words, borrowing from Plumwood's vision (2002), Gaard (2017) describes critical ecofeminism as a "relational self, dynamically interconnected with an agential nature that is far from the inert, lifeless, and mechanistic conception of property—rather it is a nature that actively constitutes earthothers" (21). In this view, land ownership is bidirectional; you own it, and it owns you. It is a place-situated epistemology, or way of knowing, that identifies colonial relationships of displacement as well as sacrificial and invisible lands, while also recognizing the communication and intentionality of ecologies. In this model, human/nature dualisms are rejected, and in their place relational and dialogic identities are inserted. Instead of rugged individualism, she sees us as part of an ecological web of connections and relationships. Her vision is that of transformation, or moving toward a nonoppressive, egalitarian world. Explicit in Gaard's argument is that we should listen to nature because if we had been listening previously, these crises would not have happened in the first place.

A radical approach to sustainability challenges us to engage the root cause of socioecological struggle, which is social in origin. The root cause of our interrelated social and ecological struggles is a human creation—an exploitative, growth-oriented economic system. This economic system is bolstered by ideational discourses and political legislation that legitimize socioecological inequity and uneven resource distribution. The solution is socioecological justice, or a radical sustainability, which requires both a transformative approach to socioecological change—dismantling of hierarchies toward total liberation—as well as a regenerative one—healing and restoring the health of people and the planet. The first step to recovery and healing is acknowledging and understanding the problem. Social sustainability studies have much to learn from environmental justice and ecofeminist scholars. In the next section, I tackle the insidious ideological manipulation and discourses that maintain and perpetuate extreme systemic socioecological inequities.

1.3. What Barriers Exist to Total Liberation?

1.3.1. The Socioeconomic Origins of Ecological Rifts

Critical environmental sociology theories, and metabolic rift theory in particular, offer a means to critique the conventional premise of sustainable development by exposing the internal contradictions of capitalism.[7] In this section, I discuss the rift-generating mechanisms of industrial capitalism. This theory highlights how the exploitation of peoples and the environment are essential characteristics of capitalism, or how the "ecological rift is, at bottom, the product of a social rift" (Foster, Clark, and York 2010, 47). The project of sustainable development, increasing human well-being and maintaining the environment while growing the economy, is undermined by its very foundation—capitalist processes. To be sure, there are examples of other political and economic systems as well as societies that have created irreparable ecological rifts and collapsed as a result (McNeill 2000; Moore 2000; Diamond 2005; Mosley 2010). However, industrial capitalism has greatly expanded rifts on an unprecedented global scale and is the current economic system advancing global rifts (Burgess, Carmona, and Kolstee 1997; McNeill 2000; Moore 2000).

According to metabolic rift theory, the original sin, so to speak, of capitalist production began with the separation of town and country, or urbanization and industrial agriculture. Marx deduced that large-scale agriculture diminishes soil nutrition and necessitates imported fertilizer in order to remain productive. Human waste is an integral aspect of soil nutrition. However, technological changes during the Industrial Revolution spurred two interrelated processes that caused the rural-to-urban migration, which disrupted the soil-nutrient cycle. Increasingly mechanized agriculture required less human labor for farming, making farming less economically viable for poorer farmers. Capitalist mass production created a demand for human labor that pulled farm workers from rural areas into urban industrial centers. This rural-to-urban transition required the exportation of food, or nutrients, away from hinterlands to urban centers. As nutrients were sent to feed growing urban populations, farmers came to rely on imported nutrients for crops. In addition, concentrations of human waste in urban centers turned into pollution rather than fertilizer for the soil. As such, the town is the consumer of nutrients while the country is the producer. This division of labor between town and country only intensified as capitalist production expanded (Foster 2000).

Metabolic rift theory further postulates that political-economic as well as sociocultural relations are in large part responsible for other ecological rifts, or environmental degradation (Foster 2000). Specifically, capitalist agriculture consists of a chain of estranged and exploitative relationships between town and country, landowner and worker, and worker and soil. Capitalist agriculture is exploitative in its relationship to the soil because the goal of large-scale production is short-term profit over long-term subsistence (for a more detailed discussion of these assertions see Foster 1999, 2000; Magdoff, Foster, and Buttel 2000; Clark and York 2005; Clement 2011b). Soil degradation occurs because a landowner has no relationship with the soil because laborers work it, and laborers have an estranged relationship to the soil because their commands come from the landowner. In this light, Marx's concept of metabolic rift describes "the material estrangement of human beings within a capitalist society from the natural conditions which formed the basis for their existence" (Foster 2000, 163). As noted above, farm workers' spatial proximity grants them direct information about soil quality; thus, democratic mechanisms may allow for better soil management (Plumwood 2002).

Also integral to Marx's concept of "rift" is the failure of capitalism "to maintain the means of reproduction" for both the soil and the worker. Reproduction in this sense means sustaining soil fertility as well as workers' health and economic viability. Without a system of "restitution," or regeneration, for labor or soil, a rift is created in both biological and ecological metabolic processes (Foster 1999, 2000). The contradiction in exploitation is that failing to maintain reproduction is economically unsustainable because production is reliant on cheap laborers—who are also consumers. If laborers are paid too little money, they cannot participate in consumption. If they cannot buy food, it undermines their ability to survive, work, and contribute to the economy—all of which capitalists depend on for profits. The need for continuous accumulation is ecologically unsustainable because commodity production is reliant on finite energy and natural resources; therefore, the system cannot exist indefinitely. However, capital has been able to extend its life through trade liberalization and globalization, which have effectively turned the developing world into the hinterlands (or country) for the developed world.

1.3.2. The Treadmill of Production in Our Modern World System

Most modern urban environments are sustained by global capitalism. Schnaiberg's (1980) treadmill of production is one of the most influential theories of environmental sociology that connects capitalism to environmental degradation. This theory posits that as capitalists seek their primary goal, the acquisition and accumulation of wealth, they continuously ramp up production processes toward that end, which manifests as relentless ecosystem exploitation and degradation. All wealth is extracted from the environment and human labor. Every social institution in society is geared toward growing that wealth, including the media, marketing, and advertising; government tax breaks and subsidies for businesses aimed at economic growth and competition in global markets; education for future workers; labor unions and workers fighting for jobs; as well as the corporations that stand to profit from growth (Barbosa 2015). Indeed, families compete in ever more elaborate displays of wealth, as cultural signals that they are keeping up with the Joneses. Some churches even preach prosperity gospels that legitimize this endless pursuit of wealth. To grow wealth, capitalists seek out cheaper labor and production processes and more easily accessible raw materials; thus, this treadmill has grave consequences for both people and the planet.

In the treadmill of production, businesses withdraw raw materials from the environment to transform them into consumer goods that are then exchanged for money on the market. Consumers devour and toss aside the remnants of these items, transforming them into garbage and pollution. In order to stay competitive, businesses must grow their operations, gaining ever larger markets and profits. They reinvest these profits to produce and sell more goods. If a business can sell its goods for less than its competitors, it is likely to increase its market share and, ultimately, run its competitors out of business. In order to produce more consumer items for less, businesses must find ways to cut labor costs. They do this by downsizing employees, coercing or enticing fewer workers to produce more; outsourcing, moving business operations to countries where workers work for less; or through automation, investing in machines that can produce more than many people can. In addition to cutting labor costs, businesses gain profits by expanding consumer markets, or tapping into latent demand through advertising, globalization, and planned or perceived obsolescence (or convincing consumers to buy a new item before the original item's useful life is over). Every stage of a product's lifecycle, from cradle (extraction) to grave (disposal), causes human and environmental consequences, or negative externalities, that may include damages to human or ecosystem health.

World systems' scholars refer to the product lifecycle as a commodity chain (Wallerstein 2004; Barbosa 2015). For example, first miners, loggers, or roughnecks extract minerals, timber, or energy from the earth; these raw materials are then shipped to refineries and factories where factory workers assemble materials; assembled items are shipped to stores where salespeople sell commodities; consumers purchase commodities; consumers then throw away packaging or undesired items; and then landfill workers bury or recycle the waste. Each step in this chain comprises a node of people and environmental goods/harms, and each node has its own set of unique problems. For instance, mining materials often releases particulate matter into the air and heavy metals that can contaminate soils and fresh water sources. Workers exposed to airborne particulates may experience respiratory problems. Raw materials are usually transported in large trucks, planes, trains, or boats that likely release greenhouse gases, which contribute to climate change. In factories, workers assembling the products are often exposed to a variety of toxins, including heavy metals or dangerous chemicals, such as flame retardants. The products are then transported again to stores where consumers purchase them, likely drive them home, and discard of them soon thereafter, given that only 1 percent of items are still in use six months after purchase (Leonard 2010). Of course, the next link in the chain is the waste dump, or the plastic island in the Pacific Ocean where fish and other sea creatures consume and are poisoned by the garbage. In some cases, the great cycle of life continues when people then eat the poisoned fish. In a globalized economy, each node of production and consumption is likely to occur in different parts of the world, requiring extensive transportation and energy. In addition, uneven relationships between nations allow more harmful production and waste disposal processes to occur in poorer nations, while affluent consumers in rich nations are better insulated from harm and benefit most from production.

Some assumptions underlying our economic system presume that free markets work most efficiently when each entity, person or corporation, is an individual who acts on rational and calculated self-interest. According to this logic, each of us is competing for limited resources, and the fittest of us will thrive by successfully attaining wealth or growing our businesses, while the weakest fail due to their own ineptitude (i.e., the weakest businesses go under). Consumers demand certain goods, and businesses compete for customers by providing the best quality products for the lowest price. Happy customers keep buying from the best businesses, creating a win-win scenario for entrepreneurs and consumers alike. As Alexander observes, the assumption that "this unrelenting individual competition" should somehow "maximize everybody's well-being in the long run, multiplying individual happiness and the wealth of nations" is dubious at best (2008, 60). Belief that rational actors create supply-demand equilibrium ignores coercive processes and power differentials between most corporations and consumers, such as targeted consumer research, marketing, and advertising. Further, this economic Darwinism prompts individuals to see each other as competitors, or as temporary coconspirators toward some utilitarian, accumulation-based end. However, once the utility of a relationship runs its course, business norms resume competitive relationships. In this way, kinship ties, friendships, and deep community connections are actually a hindrance to the most important goal, profit. This "business is business" mantra supersedes close relationships.

1.3.2.1. The Social Psychology of the Treadmill of Addiction

The structure of capitalism, that is, the competition between businesses for growth and profit no matter the cost, shapes corporate behavior, which in turn affects the relationships of individuals within the corporate system. The 2003 documentary *The Corporation: The Pathological Pursuit of Profit and Power* features prominent psychopathy researcher Robert Hare, who argues that if we look at corporations as legal persons, some of them exhibit the diagnostic criteria of psychopaths.[8] The film uses examples of corporate crimes, such as Monsanto suppressing investigative journalism that exposes the health risks of the bovine growth hormone; Bolivian protests of water privatization; and IBM's alliance with Nazi Germany to illustrate such antisocial behaviors.

Western European corporations have been integral to colonization, beginning in the mid-fifteenth century. Eventually European colonization transformed capitalism into a global economic system. But it is worth remembering that the capitalist economic structure is a mere blip in humankind's existence of more than 200,000 (possibly 300,000) years.[9] By comparison, gathering and hunting societies, and much later agricultural and pastoral societies, existed for tens of thousands of years. These societies inhabited different ecologies and engaged in diverse environmental practices and ways of knowing. However, as European colonies proliferated, so did European capitalism, Christianity, and Enlightenment philosophy. Currently, in the United States these beliefs are taken for granted as natural and the way life ought to be. Many social theorists base their assumptions about human nature on the way things are, claiming that humans are competitive and greedy by nature. I argue that the structure of our society—especially our heteropatriarchal, white supremacist, speciesist, and class-based economic system—constructs competitive, disconnected human relationships that run counter to the types of relationships found within many traditional and Indigenous communities.

I disagree that human beings are greedy and selfish, that we are naturally competitive, and seek market-based economic relations. Rather, our learned economic and social relationships affect how we treat the environment and each other. Many examples exist of different cultures from all over the world that have been more cooperative, equitable, and environmentally sustainable than ours. Those cultures are quickly disappearing as Western European ideals of modernity and progress, much like an invasive species, threaten to displace rural populations and spread consumer cultural homogeneity. This economy is based on competition and domination, which has both social and ecological consequences: social inequity and environmental degradation.

Philosophers have long theorized the nature of "man," but social theorists in the late nineteenth through the twentieth centuries sought to understand human nature in the context of industrial capitalism. Erich Fromm (1956) notes:

> The prevailing concept of man . . . is based on the structure of capitalism. In order to prove that capitalism corresponded to the natural seeds of man, one had to show that man was by nature competitive and full of mutual hostility. While economists "proved" this in terms of the insatiable desire for economic gain, and the Darwinists in terms of the biological law of the survival of the fittest, Freud came to the same result by the assumption that man is driven by a limitless desire . . . and that only the pressure of society prevented man from acting on his desires. (84–85)

In Western society, we are beholden to the tenets of private property and land ownership. In order to survive, we must sell our time and labor as a commodity to obtain enough money to pay rent or a mortgage as well as to eat food. Through globalization, this is becoming increasingly true for people around the world. Most people work for someone else, someone who owns and runs the business. Our current iterations of colonization, slavery, and labor exploitation renders workers cogs in a mass production machine.[10] Displaced from rural lands where people can produce for themselves, alienated from fulfilling work, and concentrated in urban centers, we search for meaning and connection among others who work soulless jobs. Marx called this alienation. Alexander (2008) refers to it as dislocation, which he believes has grave consequences for both individuals and society. The significance of performing labor for someone else's profit is that workers are at once alienated from themselves by selling their time, physical/mental health, and their search for purpose; alienated from others as their work is often predicated on someone else's suffering; and alienated from nature, because often their work is also predicated on the destruction of ecosystems and other creatures' autonomy.

Consider the case of slaughterhouse workers. While this example may seem extreme, most people in the United States eat meat, and these are the workers that make eating meat possible. Slaughterhouse workers experience some of the highest rates of occupational injury and psychological distress, already sacrificing their physical and mental health for affluent consumers (Leibler and Perry 2017). In fact, depression and anxiety rates among slaughterhouse workers are found to be higher than the general population (Dillard 2008; Leibler and Perry 2017). These workers tend to be more violent toward people as well. Fitzgerald, Kalof, and Dietz (2009) find "slaughterhouse employment increases total arrest rates, arrests for violent crimes, arrests for rape, and arrests for other sex offenses in comparison with other industries" (158). One worker states, "The worst thing, worse than the physical danger, is the emotional toll. . . . Pigs down on the kill floor have come up and nuzzled me like a puppy. Two minutes later I had to kill them—beat them to death with a pipe. I can't care" (Dillard 2008, 391). When they leave work, some take up drinking or other forms of escapism and may act out violently against other human beings (Dillard 2008). Alexander argues that symptoms like the ones detailed above are signs of dislocation, or an alienated person attempting to cope. Ironically, people desperate for meaning and connectivity often engage in more socially isolating behavior, as they may feel extreme shame (Alexander 2008).

As Alexander argues, dislocation can inflict anyone from any socioeconomic background, as globalized free-market society "produces mass dislocation as part of its normal functioning even during times of prosperity" (60). Despite their privilege, elites also suffer from estranged relationships. Research demonstrates a strong correlation between increasing wealth and antisocial behavior, whereby rich individuals often choose to protect their wealth over other people. This antisocial behavior presents itself in myriad forms. Sociologists Clarke and Chess (2008) coined the term "elite panic" to describe the ways in which elites react to various disasters. They fear the unfortunate masses and build militias, police and security forces, and militaries to protect their wealth, property, and power. As a current illustration of their alienation, Silicon Valley and Wall Street executives are preparing for civilization collapse by buying and stocking bunkers in remote places like New Zealand. They fear that they won't have enticing enough rewards, similar to money, through which to coerce security forces into protecting them and their resources when the economy fails (Rushkoff 2018). Alexander notes that dislocation can be thought of as "poverty of the spirit."

Fromm and Alexander seem to agree that because most alienated work pits us against each other while also dehumanizing us and our human and nonhuman community members, we experience an extreme lack of connection. For many, this manifests as insatiable desires, which our corporate media convinces us can only be filled with the consumer items that generate profit for these same corporations. As a result of social dislocation, Fromm suggests that our culture has many ways to keep us unaware of our social dislocation:

> First of all the strict routine of bureaucratized, mechanical work, which helps people to remain unaware of their most fundamental human desires, of the longing for transcendence and unity. Inasmuch as the routine alone does not succeed in this, man overcomes his unconscious despair by the routine of amusement, the passive consumption of sounds and sights offered by the amusement industry; furthermore by the satisfaction of buying ever new things, and soon exchanging them for others. . . . Man's happiness today consists of "having fun." Having fun lies in the satisfaction of consuming and "taking in" commodities, sights, food, drinks, cigarettes, people, lectures, books, movies—all are consumed, swallowed. The world is one great object for our appetite . . . we are . . . the eternally expectant ones, the hopeful ones—the eternally disappointed ones. Our character is geared

> to exchange . . . and to consume; everything, spiritual as well as material objects, becomes an object of exchange and consumption. (80–81)

Fromm (1956) concludes that "clinical facts demonstrate that men—and women—who devote their lives to unrestricted . . . satisfaction do not attain happiness, and very often suffer from severe neurotic conflicts or symptoms. The complete satisfaction of instinctual needs is not only not a basis for happiness, it does not even guarantee sanity" (85).[11] Because we are social creatures who seek connection with our human and nonhuman communities, when we are isolated, we feel profound pain. In order to mask the pain, we seek novel experiences and become addicted to a variety of things, including the consumption of consumer goods.

1.3.2.2. Treadmill of Consumption and Addiction

In her book *When Society Becomes an Addict*, Anne Schaef (1987) makes a compelling argument for what she calls the "addictive system." The addictive system, in her view, perpetuates addiction to "keep us afraid, out of touch with ourselves, and too busy to challenge the system" (13). This is necessary in any oppressive social order to maintain the hegemony of the dominant, and compliance of the subordinate, groups. She specifically refers to the white male system, but for our purposes, I speak of the elite, white supremacist, speciesist, heteropatriarchy. She characterizes the addictive system as a hologram, which she describes as each piece embodying the entire structure of the whole such that "the system is like the individual, and the individual is like the system" (37). She maintains that our addictive society surrounds us with temptation to keep us complacent and to continue business as usual. In order to function "normally," this system "calls forth addictive behaviors," by allowing individuals few choices, cultivating addictive processes, and seducing addictive behaviors (25).

The addictive system is an apparatus, and its scaffolding includes addicted individuals, codependent family members, inauthentic community relationships, and lack of paths for fulfillment. Schaef suggests that most of us play some part in keeping up the charade. Addiction processes include negative behaviors and emotions that foster self-centeredness, dishonesty, illusions of control, denial, confusion, dependency, and fear, among others. Underlying these behaviors and emotions is a complete lack of self-awareness, especially in relation to others, and usually isolation, immense self-loathing, and shame.

Addiction research points to a number of potential individual risk factors, of which I focus on the issues of availability, social dislocation, and abuse. Other risk factors that contribute to a person's abuse and/or dependency on substances or other stimuli include genetics, family history of addiction, and childhood trauma, such as neglect or physical, sexual, and emotional abuse (Dube et al. 2003; Anda et al. 2008; Khoury et al. 2010), as well as social isolation and dislocation (Alexander 2008). Addiction can manifest in a number of ways, including binge eating and anorexia or the abuse of substances, shopping, gaming, sex, the internet, and others. In this section, I explore the addictive features of our capitalist treadmill that manifest as consumption and profit accumulation. Both have to do with the never-ending drive to consume.

Schaef (1987) defines addiction as "any process over which we are powerless." She continues, it "is progressive, and it will lead to death unless we actively recover from it" (18). Alexander's (2008) definition of addiction is more complex, but he identifies what he calls "addiction3" as the most common in our modern society (62). This type includes any behavior—compulsive love-seeking, gambling, binge eating, shopping, and gaming, as well as drug and alcohol abuse—that is "overwhelming, intractable, and dangerous" (35). Alexander theorizes that addiction occurs first as a result of a lack of connection, or spatial and intimate isolation, from community and other individuals. As the illness progresses, the addict further severs social connections.

Alexander further contends "free-market society produces mass dislocation" because our "hyperindividualistic, hypercompetitive, frantic, crisis-ridden society makes most people feel socially and culturally isolated" (2010, 60). The United States, as a champion of free markets and purveyor of individualism, perfectly illustrates Alexander's thesis. Loneliness is endemic among adults in the United States. A Cigna survey found that more than half of their 20,000 respondents said they always or sometimes feel "no one knows them well" (Cigna 2018, 3). In addition, 46–47 percent of those surveyed reported "sometimes or always feeling alone and/or feeling left out" (2018, 3). Younger generations experience higher rates of loneliness and self-reported health problems than do their elders. Loneliness is more dangerous than obesity and "has the same impact on mortality as smoking fifteen cigarettes a day" (Cigna 2018 2). It also is closely related to mental illness, where in the United States about one in five American adults suffer from a mental illness (National Institute of Mental Health 2017; Cigna 2018).

A number of meta-analyses looking at changes in mental health over the last century, in North America and parts of Europe, have found increasing rates of affective and personality disorders, especially among young adults, including major depression, anxiety, and narcissism (Klerman and Weissman 1989; Twenge et al. 2010; Herbst 2011; Twenge 2015). In one such article, researchers look at birth cohorts in the United States between 1938–2007 and find "large generational increases in psychopathology" among young people, citing the primary cause as "cultural shifts toward extrinsic goals, such as materialism and status and away from intrinsic goals, such as community, meaning in life, and affiliation" (Twenge et al. 2010, 145). Similarly, another article finds "the steady erosion in social and civic engagement, interpersonal trust, and financial security" to be partially "responsible for the widespread decline in subjective well-being over the past few decades" (Herbst 2011, 773). Other researchers note that mental health problems have increased while economic security has declined (Cohen and Janicki-Deverts 2012). Research observes the connection between mental health problems and addiction, much like the ACE studies cited below. Rates of substance abuse and addiction also have seen a precipitous increase, especially in marijuana (National Institute on Drug Abuse 2015) and opioids. According to a Blue Cross Blue Shield report (2017), opioid use disorder diagnoses spiked 493 percent between 2010 to 2016 in the United States.[12] In addition, researchers in the United States have warned of a national crisis because of an alarming rise in "deaths of despair," which include suicide, alcohol, and drug overdoses (Curtin, Warner, and Hedegaard 2016; Radley, Collins, and Hayes 2019).

Mental illnesses, especially affective disorders such as anxiety, depression, and posttraumatic stress disorder (PTSD), are also closely related to childhood abuse and trauma (Chapman et al. 2004; Chapman, Dube, and Anda 2007). As a manifestation of this, adults with high adverse childhood experience (ACE) scores, which is a protocol that includes ten questions (or more in the expanded version) that measure prevalence of childhood and oppression-related trauma, are three times more likely to be prescribed psychotropic prescriptions, such as antidepressants and mood stabilizers (Anda et al. 2007). In addition to these findings, adults with high ACE scores are more likely to attempt suicide, abuse prescription drugs (Anda et al. 2007; Anda et al. 2008), smoke (Edwards et al. 2007), consume alcohol at a young age (Dube et al. 2006), and experience a range of physical health problems from autoimmune illnesses (Dube et al. 2009), lung cancer (Brown et al.

2010), and even premature death (Anda et al. 2009). In the study population of 17,000 people, about two-thirds of respondents had sustained at least one or more traumas that put them at increased risk.

A cynical view of these trends would point to the almost too convenient nature of the simultaneous rise of mood disorders and addiction, which are symptomatic of wide-scale social dislocation. Alexander (2008, 63) writes of the dark side to our addicted society:

> It is possible to dream that society will benefit from the insatiability that comes with addiction through the brilliant achievements of addictively competitive Chief Executive Officers (CEOs), the economic stimulus of addictively spending consumers, and huge government revenues from those who pour their livelihoods into slot machines and lotteries. However, such dreams pale in the face of the long-range costs of corporate and government corruption, stress diseases, family devastation, environmental destruction, and so on. Nevertheless, corporations compete by systematically encouraging addictive consumption in their customers and addictive work habits in their employees, thus acting as "pushers" for the most common addictive habits of our times. Their incessant advertising lulls us in our pallid dream.

Consumption is the other side of production, and it must continue to grow in order to metabolize production, hence the treadmill of consumption. "Consumer culture" sounds so innocuous and does not adequately explain the set of social relations that underpin it. Seamlessly, capitalism creates the conditions necessary to sustain it. It alienates and isolates once socially connected beings, and then sells the bereaved temporary fixes. As the social isolation worsens, capital has secured lifelong customers. Yet here we stand, confronting catastrophic climate change, but in sheer denial of the dangers we face, we search ever harder for fossil-fuel deposits. What else can we call this but a treadmill of addiction? The authors above argue that our addictive system, capitalism—and I'd add all other interconnected oppressive structures, such as speciesism, heterosexism, and white supremacy—is able to maintain destructive practices because it creates trauma for those it marginalizes and exploits; it contrives dependency and social isolation; it enables addiction; it facilitates codependent addicted relationships; and it manufactures compliance through palliatives and marketing that preys on insecurity, fears, and unmet desires.

The concept of addiction manifests on every scale, from the individual to the global, and propels the treadmills of production (the forever growing of our economy and GDP), and consumption (the endless pursuit of consumer goods), that necessitate the treadmill of destruction (military force). Our addictions to production as a society and consumption as individuals necessitates military force to seize resources from elsewhere and to protect the resources we have. Each of these treadmills—production, consumption, and destruction—is a driver of climate change, ecosystem destruction, and human suffering. They are killing us and our planet. Of course, the richest among us are the most protected from this suffering, because they benefit the most and endure the fewest consequences, which is how this treadmill is able to continue. Elites also have power—with all the money, politicians, and media on their side—to coerce and/or con the rest of us into submission.

1.3.2.3. Ideological Manipulation

Industries pair with public relations firms, think tanks, and political campaigns to construct a master narrative that is meant to convince us that the way things are is the way they ought to be. Many examples of manipulation designed to keep the public unquestioning and workers complacent exist. Environmental skepticism, for instance, is a conservative, elite-driven movement mostly based in the United States intended to spread environmental skepticism narratives that deny the seriousness of problems, dismiss scientific evidence, and claim that environmental protection policies threaten "progress" (Jacques, Dunlap, and Freeman 2008). Recently, ExxonMobil provoked controversy when researchers exposed the inconsistencies between ExxonMobil's own scientific research and the information they presented to the public regarding climate change. In particular, most of their research acknowledged anthropogenic climate change and blamed the burning of fossil fuels for rising atmospheric concentrations of CO_2. However, in advertorials (advertising designed to look like editorials) in newspapers such as the *New York Times*, most of their writing about climate change expressed doubt about its causes (Supran and Oreskes 2017). ExxonMobil also has funded climate denial think tanks and political lobbyists opposing climate policy (Jacques, Dunlap, and Freeman 2008; Farrell 2015). So far, these attempts by corporations with vested interests to create doubt have actually proven successful. Although climate change and its human causes have been thoroughly and scientifically documented since before the 1970s, even today only about half of Americans believe that climate change is mostly human caused (Leiserowitz et al. 2015).

The lengths an industry will take to manipulate its workers and the public it "serves" to protect its own interests can be best understood through another example. I turn to the coal industry in the Southern Appalachian region of the United States.

Capitalism at once causes the problems of dispossession and social dislocation and sells back the solutions to the dispossessed, creating "consumers" of land, food, water, and escapism. It creates dependence by stealing people's land and natural wealth, displacing them, and making them unable to subsist for themselves. Then either by keeping them landless squatters, temporarily loaning them land, or selling unproductive land back to them and coercing them into purchasing food, people are forced into servitude, slavery, and exploitative labor. Thus, most people need to labor for someone else in order to sustain themselves, and they fight to maintain that labor, even when the work itself is slowly poisoning and killing them. Without the work, they have no way to pay for food or homes, medicines to combat the poisons, and analgesics to combat profound existential pain.

The history and present state of coal mining in Appalachia clearly demonstrate this dynamic. Prior to coal mining, Appalachians were largely subsistence farmers. When prospectors discovered coal, they duped locals into unfair land exchanges or pushed landowners unwilling to sell into legal traps, coercing them into selling land and mineral rights. As a result, the coal industry now owns between 70–90 percent of the land in West Virginia. Locals whose ancestors were once self-sufficient now need to work for food and shelter, and eagerly take employment in the coal mines because there are so few decently paying jobs in Appalachia. As coal mining has become increasingly mechanized, people have lost jobs even as coal extraction increases. Concurrently, many people have fallen ill or lost property as a result of coal mining practices (Bell and York 2010). In order to maintain local hegemony despite declining employment, the coal industry needed to take action to keep local people from fighting back.

Sociologists Bell and York (2010) present a study of how the coal industry actively constructed the economic identity of West Virginia into the twenty-first century. They claim that through ideological manipulation, coal elites maintained power and legitimacy by hiring a public relations firm, funding an Astroturf organization, and sponsoring local events. As a form of ideological manipulation, the coal industry framed itself as the "backbone of local and regional economies." They pursued this framing *despite* the significant reduction of mining jobs in the last half century, down to about 5 percent of the total labor force, and the general decline in coal's contribution to the local gross state product (GSP), down to about 7 percent of GSP and trailing at least four other employment sectors (Bell and York 2010, 112). Bell and York argue that as the coal industry's economic contribution waned, they feared a legitimation crisis, "where the public comes to reject ideological justifications for the prevailing social system," and in order to maintain hegemony in the region, they needed to convince people of their continued importance to the Appalachian economy (117).

To do this, coal elites used multiple discursive strategies, including diversionary framing and colonizing the Appalachian "lifeworld," the locals' everyday lived experiences and culture, by commandeering local cultural icons, constructing the appearance of ubiquitous community support, and sponsoring local sporting and cultural events. Diversionary framing consists of discourses "which sidestep the specific environmental complaints of critics, refocus the debate on larger macro issues tied to US energy production and economic growth, and discredit the legitimacy and rationality of groups opposed" to more extraction or production (Ladd 2014, 301). Other tactics included the coal industry's establishment of an Astroturf organization called Friends of Coal to speak on their behalf. They gave away free "Friends of Coal" hats, pins, shirts, and so forth to foster the appearance of universal community support. Public relations campaigns evoked traditional gender roles to appeal to men's nostalgic identities as breadwinners, as women historically cared for the home when men went into the mines. They incorporated local heroes who embody the archetypal West Virginian man to defend coal in ad campaigns. These male icons included successful retired football coaches, an Air Force general, and a professional bass fisherman. These insidious public relations practices created the illusion of continuous coal and male dominance in the region, despite all the evidence to the contrary (Bell and York 2010). Indeed, other energy-related resource extraction industries use similar tactics, such as uranium mining companies in the Four Corners region of the United States (Malin 2015).

1.3.2.4. Denialism

Both ideological manipulation and addiction rely on strong and unyielding denial. In the North, our main addiction is insatiable natural resource consumption, which is causing devastating social and ecological deterioration.

Alexander (2008) sees this as a means of adapting to painfully severe social dislocation, or a lack of social integration that is endemic of a free-market economy. He reasons, "at the same time that the free market dislocates people, it proffers pseudosolutions for the misery of dislocation" (250). He observes that corporations push consumer goods to temporarily fill the void and actively normalize shopaholic tendencies. Palliatives include mostly consumer items, such as palatial homes, designer clothes, and exotic foods. Indeed, consumption of luxury items has dramatically increased during the last half century, as modern consumers consume twice as much as they did fifty years ago (Leonard 2010). This excessive consumption has significant socioecological costs, which Alexander notes we push to the "periphery of consciousness, as in addictive denial" (250). Integral to this denial is that individuals, communities, and governments actively ignore the harm an addiction is causing, harms such as climate change and enduring poverty (Alexander 2008). Schaef (1987) similarly maintains that denial holds an important function within an addictive system, and everyone plays their part. She further notes that it is "when we . . . start trusting our own perceptions, we become a threat" (81).

Norgaard (2011, 2015) examines the cultural and political-economic contexts of climate change denial, or what she terms as the "social organization of denial," which comes out in norms of conversation, attention, and emotion to create an "insidious form of social control" (2015, 253). To dismiss disturbing information and avoid negative emotions like fear, guilt, and helplessness, people use avoidance strategies in conversation, such as changing the topic, telling jokes, trying not to think about it, or just failing to bring it up, even in political contexts. She invokes Lifton when she describes how "fear inhibits our ability to break through 'illusions' to 'awareness'" (2015, 257). Maintaining these illusions requires a great deal of "numbing" in everyday life to remain comfortable while attempting to suppress the tension and anxiety that comes with knowing that our survival as a species is at stake. Some devices people use to remain comfortable are structurally supported literal denial, American exceptionalism, anti-intellectualism, and individualism. Specifically, under the George W. Bush administration, officials suppressed reports, falsified documents, and officially stated that climate change needed further investigation (2011). In the United States, skeptics attack climate scientists by calling them "alarmist," "junk scientists," and "against progress." American exceptionalism places "the American way of life" as beyond examination, and anything that challenges it, such as asking Americans to decrease their energy consumption, is undemocratic (2011). Anti-intellectualism is firmly rooted in American political culture. Americans challenge science as a "legitimate epistemology in the public sphere," and question science as a tool for gathering collective information and informing policy (2011, 203). Aside from questioning climate change, some in the American public question scientific understandings of evolution and vaccinations. Finally, a strong American ethos of individualism combines with a strong distrust of the political system that serves to disempower people, causing inaction in the face of large-scale social problems, such as climate change, that require large-scale collective solutions in the political sphere (Norgaard 2011).

1.3.2.5. Individualism and Social Isolation as Social Control

Hyper-individualism is a Western narrative, proliferated in the United States, that conjures images of white settlers conquering the "wild" West while surviving off the land alone with their families and protecting themselves against the elements. In the United States, values of autonomy, independence, resourcefulness, self-reliance, and success are upheld as noble characteristics (Callero 2018). More recently, globalization and the expansion of neoliberalism has spread the culture of individualism across the globe. While individuals may be freer in some senses regarding personal choices, they become politically ineffective, especially if they have little wealth, because their power comes from numbers. Collective action is the only available means that most people have to create structural and cultural change. Hyper-individualization also is in part responsible for, and legitimizes, social isolation and dislocation (Norgaard 2011). As noted above in the research on loneliness, most people don't feel like they have much in the way of community.

The irony is that the elite class is an organized group of individuals that have organizational power as well as economic, political, rhetorical, and militarized power (Domhoff 2009). It seems all too convenient that political leaders would extol individualism, given how powerless and politically ineffective it keeps the rest of us, which brings me to my last point about systemic abuse. Individualization, political disenfranchisement, and marginalization are all structural forms of social isolation. In an abusive relationship, abusers generally socially isolate their victim. This allows the abuser to control the narrative and the victim's actions. It also hides the abuse from everyone else. Significantly, the victim has no one to turn to for help because they are estranged from all of their friends and family. This isolation allows the abuse to continue indefinitely and renders the victim powerless (Wiener 2017). Fortunately, this book offers strategies on how to take the power back!

1.4. Toward Socioecological Transformation and Regeneration, or Radical Sustainability

If we can make proper diagnoses, then we can evaluate potential remedies. If what we live in is a systemic treadmill of addiction, then the potential remedies would include forms of psychosocial reintegration, such as counseling aimed at individual and group psychological healing to unlearn and recover from systemic oppressions (related to class, race, gender, sexuality, ability, and species, among others). We would need to reprioritize actual emotional, physical, spiritual, and creative needs versus excessive material desires. Moreover, we would need to reorganize production systems toward collectives and revalue economies based on happiness, health, and other quality of life measures. Also integral to radical sustainability is regaining connectedness to local natural environments so that we can see signs of problems before they become unmanageable. Some socioecological theories can help us conceptualize the healing path toward radical sustainability—as noted above, radical sustainability emerges from self-sufficient communities that are premised on ecological principles. My cases are presented here to help fill in some theoretical gaps.

Metabolic rift theory paired with insights from social sustainability and the total liberation framework offer pathways for potential solutions. Specifically, metabolic rift offers a vision for how to simultaneously regenerate soil nutrition and transform social relations to create nonhierarchical, collaborative food production systems.

1.4.1. Restitution

Marx argues that a rational agricultural system is based on restitution. Restitution, or what I refer to as the process toward a radical sustainability—social and ecological transformation and regeneration—is the process through which the metabolic rift is mended. This concept originates from Liebig's work on soil nutrition, and, more specifically, it means "giving back to the fields the conditions of their fertility" to ensure "the permanence" of soil fertility (Foster 2000, 153). Informed by Foster's work, I use restitution more broadly to mean restoring any metabolic processes in which capitalism has created a rift, including humans' relationship to nature, human economic relations between each other, and the relationship between town and country. All three of these processes are interrelated. I am most interested in understanding the social components of restitution.

According to Foster's (2000) interpretation of Marx, in order for restitution to occur, the associated producers must rationally plan agriculture to eliminate the antagonism between town and country. Restitution understood this way is a process toward social and ecological transformation and regeneration, or radical sustainability. The associated producers are a collective of laborers and free farmers who own in common their means of production, such as a cooperative. Foster specifies that eliminating the antagonism between town and country includes three processes: first, the integration between industrial and agricultural production; second, a more even dispersal of the population between town and country; and finally, the return of waste from both human and industrial production and consumption to the soil as nutrients (169). However, Foster does not go into detail about what each step toward restitution actually looks like, nor does he detail the social relations between associated producers. I aim to further the theoretical understanding of restitution by incorporating insights from theories above on socioecological sustainability as well as by showing these processes in everyday interaction at a Cuban urban farm and an ecovillage in the United States. From the rich detail my cases provide, we can see what struggles people face trying to engage in direct democracy and community activities while living off the land.

Radical sustainability includes theories of restitution and social sustainability that are built on a total liberation framework and include psychosocial reintegration. These insights include acknowledging that all oppression is linked and pushing for greater equity, democracy, and well-being. I further argue that to realize transformation and regeneration, also described as radical sustainability in this chapter, we also must work toward psychosocial reintegration. Each dimension of social sustainability feeds into the other, because equity is important for both meaningful democratic participation as well as access to the resources that fulfill human needs. Similarly, fulfillment of basic needs is requisite for democratic engagement that could affect equitable access to resources. However, in the United States there are cognitive and emotional barriers to realizing social sustainability—reinforced and perpetuated by our elite, white supremacist, speciesist, and heteropatriarchal political, economic, and cultural institutions. These barriers amount to a collective amnesia and denial of our current and historical connections to colonization and slavery as well as exploitative relationships toward reproductive labor, productive labor, and ecosystems. In order to fully obtain restitution as a collective, we also need to do the emotional work of acknowledging, discussing, and admitting fault in contributing to these oppressive structures. Only through this recognition and engagement can we begin to reconnect and collectively heal. We must simultaneously work on psychosocial reintegration while attempting to achieve meaningful democratic participation, equity, and well-being.

1.4.2. Psychosocial Integration and Community Efficacy

The authors who engage systemic addiction and denial offer some insights into possible roads to recovery. Some highlights include naming the interrelated problems of environmental and social harm, taking responsibility for our part in it, having open and honest dialogue about negative emotions, actively listening to build community, considering hopeful alternatives, and collective organizing for action. Suggestions toward recovery and healing address every scale of action, from the individual to the global. I outline these insights here.

To combat socially organized denial, Norgaard (2009) argues that we must have honest conversations about what changes are needed, where it would come from, and what the benefits are. We also need to calculate the costs of ignoring these problems. Providing hopeful examples in such conversations is important to bolster positive feelings, such as inspiration. She adds that we must:

> build on positive stories of success; create a sense of community by building on the knowledge that individuals are part of a larger committed and motivated citizenry; highlight the caring which IS present in order to build a sense of pride and community; provide specific opportunities to engage in realistic actions; and suggestions should be realistic in order to be deemed credible. (47)

Norgaard maintains that suggested actions "should highlight doable changes at the same time as they encourage significant action. In order to elicit a response, people must be given not only information, but something to do" (47). It is my hope that this book helps to move along a conversation about doable actions.

Schaef (1987) calls for a shift from the destructive, nonliving addictive system to the "living process system," which is life-supporting and life-producing. She notes that in order to for us to forge a path to recovery, we must first name and acknowledge the problem. That is, addictive people and systems are dishonest, controlling, and self-centered, which adequately describes our political economy. Naming our society's addiction to the unsustainable use of energy and environmental resources as well as the exploitative social relationships these addictions are predicated on is necessary. If consumer goods reflected their true costs[13] —the real expenses of labor reproduction, health care, flourishing ecosystems, and restoration—then most of us would feel compelled (or be able to afford) to consume only to sufficiency. However, that is not the case in the United States under our current economic regime. We also would need to learn to understand our own feelings as well as the feelings of others, which proves difficult in a society where we are taught to suppress emotions and are constantly gaslighted—or made to question our own realities (I discuss gaslighting in depth in chapter 4). Labor exploitation is real and is painful; heterosexism and racism painfully penetrate our internal worlds and have socially isolating consequences; the destruction of our natural world is also painful. Honestly acknowledging and discussing these feelings is part of the recovery process, as is taking responsibility for one's part in these oppressive structures. This entails acknowledging how we benefit from, and act to preserve, certain privileges related to gender, race, class, or species. This is part of finding the power in our powerlessness. With eyes wide open in constant self-reflection and honesty, we can continue the recovery process toward alternative pathways that honor ourselves, each other, and our earthothers.

Alexander's (2008) solution is concerted social action toward "domesticating" the free-market society. He suggests a mass movement that will control the free market indirectly by reducing dislocation. He proposes we join our neighbors in social action (363). Some things for us to overcome at the national level include disrupting the devastating effects of our mass media that indoctrinate us into hypercapitalism, competition, and consumption while rarely acknowledging the consequences. Reclaiming land is necessary and can be done by both supporting Indigenous efforts at land reclamation and fighting the displacing tendencies of land speculation and development, which cause gentrification and only serve to disintegrate communities and further dislocation. Changing drug laws and treatment are important because punitive measures and mass incarceration have caused more problems than they've solved and have created politically and economically disenfranchised and marginalized community members. Reviving community arts binds communities by bolstering shared identities and meaning. Embracing multiculturalism increases communities' capacity, ingenuity, and resilience. Finally, reclaiming spirituality helps communities deny forms of religion that promote hypercapitalist ideologies, such as televangelists and prosperity gospels. On a global scale, he calls for movements that push for the heavy taxation of market speculation, renegotiated trade agreements that allow cooperatives to challenge corporations, and mass boycotts and divestment in destructive industries. I am less convinced that we can tame the free market, especially considering the monetary power corporations wield and their ability to fragment us and render us ineffectual through ideological manipulation, land dispossession, isolation, and addictive consumption as well as through classism, heterosexism, racism, and xenophobia. However, if we put life-affirming systems into place, we would in fact inhabit an entirely new system.

In sum, integral to countering denial and initiating action includes talking honestly about the issues and negative emotions; evaluating alternatives that are possible, useful, and beneficial; and taking responsibility for action. In addition, open dialogue about how our current system has served to oppress the vast majority of us and how we plan to combat these forms of oppression into the future is integral. Importantly, this requires that we fully and actively listen to each other's stories, even and especially if they make us feel uncomfortable. If we are to honestly confront oppressive and destructive forces, we must sit with the discomfort. However, listening to others' suffering should inform our action, not stall it. These conversations are important for personal empowerment and self-efficacy as well as community engagement. Together, we can create a counternarrative of self-efficacy and self-sufficiency from corporations and the state. From there, we can build coalitional mass movements for social change. Guiding questions should include: how do we have real solidarity? Given the vast and grave environmental threats we confront, how do we cooperate and build capacity rather than disintegrate as a society into civil war? How do we build truly resilient communities?

Indeed, few social transformations have ever occurred without mass movements. Taken together, the suggestions toward a path to recovery are about reclaiming our rights to space, healthy environments, and community. Each step sketched above requires a great deal of time and energy. The enormousness of this process is not lost on me, nor the movement actors working toward these goals. Even so, there are movements, including the ecovillage community and Cuban urban farm detailed in this book, that demonstrate how some aspects of this process can be done. Indeed, in Cuba, community interaction and engagement are culturally entrenched, whereas people at the ecovillage literally create the space for community.

2

Grassroots Sustainability in a Concrete Landscape

An Urban Ecovillage in the Pacific Northwest

Grassroots activism is a type of political organizing that begins with ordinary people who want to bring about some kind of larger-scale social change. The term evokes images of the indispensable, underlying root system that forms the base of the grass plant, supporting it and moving resources such as water and nutrients to the rest of the plant. It is a bottom-up, or community created, movement, as opposed to top-down, or government instituted, reform. I love the symbolism of grass sprouting through the concrete committed to containing, suppressing, and stifling its growth. Concrete is a perfect metaphor for any constraining structure, especially political and economic social relations that maintain social inequities and environmental problems. There are different kinds of grassroots organizations. Some commit to confrontational tactics or direct action, some seek reform using lawsuits as a primary tactic, some seek revolutionary change in a social system, and some people seek to live the change they wish to see in prefigurative groups. The case I focus on in this chapter is an example of a grassroots organization in an urban community seeking to "be the change" but also engaging in some reformist political work locally. Asaṅga is an urban ecovillage, or environmentally focused intentional community, located in a small city in the Pacific Northwest United States.[1]

Grassroots social movements, especially in their nascent phases, have to work within the confines of sometimes very challenging political-economic contexts as they experiment with green innovations and practices. In this chapter, I explore these dynamics with my observations of Asaṅga, a community where residents experiment with multifaceted, radically sustainable, autonomous development. These ecovillagers perform a more holistic form of sustainability that encompasses equity, psychosocial (re)integration, democracy, and the environment. While economic viability is crucial, many

Surviving Collapse. Christina Ergas, Oxford University Press. © Oxford University Press 2021.
DOI: 10.1093/oso/9780197544099.003.0003

ecovillagers prefer to work just enough to get by in order to pursue other more fulfilling endeavors. Their socioecological experiment includes living intentionally in a small, tight-knit community; practicing consensus decision-making as a direct form of democracy; attending to individuals' emotional and community involvement needs; implementing permaculture as a means to organically grow food for community consumption; and cooperating in work parties to maintain the communal spaces. To begin with, I paint a portrait of the ecovillage.

2.1. Sprouting Change in the City

I happened upon Asaṅga in the summer of 2004, while I was on a six-month volunteering assignment for a Catholic Worker establishment that sheltered unhoused young mothers. My visit at Asaṅga was my first experience with an environmentally conscious community, and I walked away inspired to imagine life beyond the cookie-cutter suburbia that I was familiar with. I returned three years later to conduct research on sustainable living. Fortunately, two of the three community's property owners and residents, Ralph and Emily, still recognized my face and met with the other villagers to reassure them that my presence would not be intrusive. I spent more than two months in 2007 visiting and living in the community; interviewing, observing, and participating in community activities; and engaging villagers in discussion. In exchange for my sleeping arrangement, a futon mattress in a teenager's living room, I moved compost, cleaned rabbit cages, swept the kitchen, and became absorbed in some individuals' environmental awareness projects. After my stay, I continued to visit the community about once a month for the next six months.

While in the community I interviewed twenty-four ecovillagers, including twenty-three of the twenty-seven adults. I also interviewed a woman who had lived there for a total of three years and had since moved. The ecovillage population is constantly changing but seems to remain consistently multigenerational. My interviewees' ages ranged from nineteen to seventy-seven, with an average age of thirty-six. Fifteen interviewees self-identified as female, and nine self-identified as male. Every interviewee was white, mostly Western European ethnics, a few Eastern Europeans, and a few individuals who claimed to have small parts of Native American ancestry. Of the twenty-four people I spoke with, twelve had lived there for at least a year.

Asaṅga is embedded within a unique neighborhood characterized by overgrown lawns, lavish fruit trees, herb garden-lined sidewalks, and houses with colorfully painted wooden frames, all situated within a conventional grid of square blocks. The ecovillage sits on five parcels of land, approximately an acre, and takes up about half a neighborhood block. The layout of the village is elliptical, with the longest distance stretching east to west. From the street, it is difficult to discern that the village consists of much of anything. It is surrounded from the east by a wooden fence that wraps around the corner within the confines of the sidewalk. As the fence moves west, it turns to cob (an earthen building material), embedded with expressive, ceramic mosaics beyond the south-facing cinderblock-paved driveway. Within the walls, the dwellings follow a similar path, situated around the perimeter of the five parcels. The assorted lodgings vary from small wooden cottages to earthen apartments (built with straw bale or other eco-friendly materials) to individual-sized geodesic domes made of weather-protected cardboard. Many materials that make these homes were scavenged from city waste, including abandoned building sites and dumpsters.

In the center of the village is a concrete tile driveway, which I helped build, decorated with leaf imprints and small mosaics. This driveway is the home to a small purple car and a small truck typically adorned with long wooden planks, tools, and several five-gallon buckets. At the end of the driveway, the woodshop-cum-garage supports a home where Ralph and Emily live. The driveway is often the site of work parties, which consist of community members working collaboratively to beautify the property or build useful and decorative additions. This is where residents combine artistic creativity and ecological design with utility to create a variety of domestic eco-tools. Expansive vegetable and herb gardens flank both sides of the driveway, and fruit trees grow throughout the village. On a summer walk through the ecovillage, one will likely encounter many earthy aromas, including ripening tomatoes, a variety of herbs, alpaca manure, and the nearby compost heap.

On a typical summer day during my fieldwork in the ecovillage, villagers began to wake up about an hour after sunrise. There was often chatter in the morning as some people watered their gardens or got together to make breakfast. Breakfast usually consisted of fresh vegetables from the garden, "dumpstered" bread,[2] goat's milk from a friend's farm, and/or eggs from the communal chicken coop. People discussed their plans for the day. Most would leave to either work or play in the city while a few stayed on site to maintain the property. Most had jobs (all but three of my interviewees) that were in line with sustainability goals. One woman was a nanny who worked fifteen hours a week and was able to bring her three-year-old daughter along with her to work. The arrangement allowed her to raise her daughter with her environmental consciousness. Another woman, an acupuncturist, worked one to two hours a week. A few people were "integrative intimacy coaches" trained in nonviolent communication (NVC), helping others understand their personal feelings and needs. There were some permaculture teachers, natural builders and carpenters, and gardeners. The rest of the individuals worked maintaining the property, trading their work for a place to sleep.

The village usually began to buzz again around five in the afternoon when people returned from their jobs or from their bicycle journeys to make dinner and work on community projects. At the time I stayed there, residents were getting ready for an eco-fair off the property, where participants from all around the state would take on projects aimed at sustainable practices. During the afternoon and into the early evening, the woodshop remained open while villagers worked on their projects. Ralph was busy making wooden fruit driers that used solar heat and air instead of electricity. Huck, a longtime resident, had a crew of young women working with him to build icosahedral huts made from cardboards, plastics, and other random city waste materials. A young man named Ears was also working with a young woman and her tent partner to build an educational sustainability sunflower wheel, backed by plywood, that would provide information on how to achieve more sustainable living in daily life. Emily had about half of the ecovillagers rehearsing a play she wrote and directed about sustainability. This activity usually continued until around ten at night, when people began to retire to their respective beds.

2.2. The Ecovillage as Sustainable Grassroots Innovation

Ecovillages, a specific form of intentional community, are relatively new phenomena. An intentional community is defined as a group of people, usually at least five individuals, including some not related by blood, marriage, or adoption, "who have chosen to live together with a common purpose, working cooperatively to create a lifestyle that reflects their shared core values" (Kozeny 1995, 18; Smith 2002). Communitarians, or individuals who live in intentional communities, may inhabit a suburban home, an urban neighborhood, or rural land in a single residence or in a "cluster of dwellings"

(Kozeny 1995, 18). Intentional communities encompass collectives spanning from religious communes to urban housing cooperatives. Ecovillages are just one type (Herring 2002; Smith 2002).

Robert Gilman (1991) formally coined the term "ecovillage" in reference to combining ecological design with a community-building design (10). As the prefix "eco" implies, ecovillages are purposeful attempts at sustainable environmental communal living. Ecovillagers may use natural-building techniques, constructing buildings made from earthen materials, and situate housing units around green space for subsistence gardening. Villages are purposefully laid out to maximize environmental efficiency and to foster community interaction (Gilman 1991; Kirby 2004; Litfin 2014). The number of ecovillages has proliferated since their inception, and ecovillages have become global phenomena. In 1990, there were eight ecovillages recorded in an intentional community listing (Smith 2002); by 2017 one source listed 1065 such communities in the United States alone (Fellowship of Intentional Communities 2017). The directories are not complete, because many communities refuse to be included; thus, a definitive number of communities is difficult to calculate.

Because of the rapid growth of ecovillages, ecovillagers' outspoken critiques of capitalist accumulation and consumerism, and their unconventional living arrangements (Walker 2005; Litfin 2014), some scholars define these networked groups of individuals as a burgeoning social movement (Schehr 1997; Kirby 2004; Ergas 2010; Ergas and Clement 2016). While some ecovillages constitute prefigurative social movements in that they only attempt to live their political vision of the future,[3] some urban ecovillagers engage in more confrontational social organizing (Boggs 1977). Specifically, at the ecovillage I visited, several members are involved in a project to design city ecovillage zones and codes, which would allow things such as multiple family occupancy and the reuse of gray water.

Seyfang and Smith (2007) argue that grassroots social movements often create and experiment with niche innovations. These authors use the term "grassroots innovations to describe networks of activists and organizations generating novel bottom-up solutions for sustainable development" (585). They further suggest that "in contrast to mainstream business greening, grassroots initiatives operate in civil society arenas and involve committed activists experimenting with social innovations as well as using greener technologies" (585). An example of the difference between grassroots innovations and more mainstream business-as-usual "green branding" is the difference between community-supported agriculture (CSA) and large-scale organic agriculture. CSAs are generally run by local farmers who cut out middleman supermarkets, making their produce local and less resource-intensive. On the other hand, large-scale organic agriculture might guarantee that certain pesticides and fertilizers aren't used in their production, but they may continue destructive practices like deforesting, spraying "natural" pesticides that pollute rivers, or exporting produce to far-off destinations. Seyfang (2010) contends that in order for grassroots innovations to diffuse into mainstream practices they must be replicable, scalable, and translatable to mainstream settings; thus, the market must be able to appropriate, or coopt, them under the current economic context.

Ecovillages exhibit a grassroots attempt to create far-reaching, niche sustainability innovations in living arrangements, agricultural techniques, consumption behaviors, decision-making alternatives, and cooperative working relationships. They define sustainability in a radically different way than, and in opposition to, the neoliberal sustainable development paradigm. They see each innovation as part of a whole system that functions together to create more sustainable relationships between community members themselves and their natural world.

I investigate the ecovillage of Asaṅga to illuminate ecovillagers' innovations toward restoration within a small city by utilizing Foster's (2000) outline for metabolic restitution as well as insights from social sustainability theories. Taken together, insights from these theories and practices outlined below constitute a radical sustainability that is at once socioecologically regenerative and transformative.

2.3. Asaṅga's Political-Economic Context

Asaṅga's context influences and shapes its version of sustainability. Thus, it's worth mentioning that Asaṅga is situated within a city in the United States, which is of course a capitalist country and democratic republic. Theories of metabolic rift and social sustainability, outlined in chapter 1, are based on capitalist economies and offer means of conceptualizing the processes of transformation and regeneration, or radical sustainability, in Asaṅga. As previously explained, metabolic rift theory elucidates that our social and economic relations are in large part responsible for social and ecological rifts, such as social inequity and environmental degradation (Foster 2000). In this section, I revisit aspects of radical sustainability, the theory of restitution, and total liberation. I outline steps toward restitution, illustrating not only the possibilities for it but also challenges to it, by looking at the case of the urban ecovillage.

Marx argued that a "rational," or sustainable, agricultural system is based on restitution. Restitution is the process through which a metabolic rift, such as rifts in the soil-nutrient cycle or in community relationships, is mended. In order for restitution to occur, the associated producers must rationally plan agriculture to eliminate the antagonism between town and country. The associated producers (from here on referred to as cooperative laborers), are a collective of laborers and free farmers who own in common their means of production. The antagonism between town and country is related to the flow of nutrients from rural soils to urban centers where waste becomes garbage rather than fertilizer. Foster specifies that eliminating the antagonism between town and country includes three processes: first, the integration between industrial and agricultural production; second, a more even dispersal of the population between town and country; and, finally, the return of waste from both human and industrial production and consumption to the soil as nutrients (Foster 2000, 169). Restitution for workers and the soil must occur for a radical sustainability to exist.

If, as Foster argues, exploitative human relations are at the center of anthropogenic ecological degradation, then a theory of restitution should incorporate insights from social sustainability that are built on a total liberation framework. These insights include nonhierarchical relationships and suggest that all oppression is linked. Further, greater equity, democracy, and well-being require work toward psychosocial integration, which includes both individual autonomy and social belonging, discussed at length in chapter 1. Such integration involves communities collectively doing the emotional labor of acknowledging, discussing, and admitting fault in contributing to oppressive social structures. Building community connections is crucial for psychosocial reintegration as well as meaningful democratic participation, equity, and well-being. In the sections that follow, I argue that ecovillagers attempt grassroots innovations in cooperative labor practices and rational agriculture, thereby somewhat alleviating the antagonism between town and country in a capitalist economic context.

2.4. A Model of Sustainability

To explore the process of restitution toward radical sustainability, or socio-ecological transformation and regeneration, at Asaṅga, I attempt to elaborate on three metabolic rift concepts with real examples from ecovillagers' everyday lives. These concepts are cooperative labor, rational agriculture, and eliminating the antagonism between town and country. In addition, I incorporate insights from social sustainability literature by discussing community decision-making practices, sharing, and caring relationships. Importantly, social and ecological practices significantly overlap in this ecovillage. As a roadmap for this section, following each of the three restitution concepts are subheadings connecting themes from my data and the social sustainability literature. Specifically, under cooperative labor, I explore consensus decision-making, work parties, land ownership, and gardening. Under rational agriculture, I discuss the components of permaculture, which are caring for the earth, caring for the people, and sharing the surplus. Finally, under eliminating the antagonism between town and country, I engage the components of restitution, such as even dispersal of the population, integration of industry and agriculture, and restoring soil nutrients. I end this section by discussing the challenges residents continue to face as a result of living in a capitalist society that constrains their ability to fully realize nonexploitative human and nature relations.

2.4.1. Cooperative Labor

Suitably, an ecovillage is a type of commune, and it was Marx's study of communes that inspired his idea of cooperative labor. Marx defines free farmers, or cooperative farm labor, as individuals who own their labor and collectively utilize the land for food production. In a cooperative labor system of production, there is no chain of exploitation, and workers can freely employ directly democratic forms of decision-making. Free farmers also can more easily establish a relationship with the soil, compared with laborers in a capitalist system of agriculture, because they take no orders from an owner. Spatial proximity and worker autonomy allow laborers to at once work their own soil, take note of potential problems, respond to ecological stressors, and reap the benefits of their harvest (Foster 2000; Plumwood 2002; Gaard 2017). Thus, their work is creative, as they are sowing and growing life, and the life they grow provides them nourishment. At the ecovillage, villagers attempt to work cooperatively by making decisions through consensus, having village-wide work parties to clean up the property, and growing their own vegetables for personal and collective consumption. Villagers' main obstacle to becoming free farmers is landownership, which is a feature of the tension between the use value, utility for residents, and the exchange value, or market worth, of the city (Harvey 1982; Logan and Molotch 1987). I describe this issue in more depth below.

2.4.1.1. Consensus Decision-Making

Ecovillagers attempt to relate to each other nonhierarchically by convening regular village meetings and practicing consensus decision-making. Meetings are held to discuss community issues, grievances, and collective solutions, and consensus is practiced at meetings to ensure that everyone's voices are heard. Consensus is a form of decision-making that requires groups to come to solutions that everyone in the group can agree on. In some cases, groups decide to come to modified consensus because not everyone can agree on a solution. In these situations, some individuals may choose to set aside their personal feelings to allow the group to come to a decision. However, if someone strongly disagrees with a proposal, they may choose to block the proposal entirely, forcing the group to think through other options. Emily, an older resident, describes how issues are addressed at the ecovillage:

> The way that works is we just get a notice out about the topic, when and where the meeting is, and anyone who cares just shows up, expresses their opinion, or hopefully, runs by consensus. Which, I'm really surprised at how well it has worked. As soon as we started using consensus, I thought we would get bogged down by all the details like what color the paint should be, but it hasn't worked out that way. People are really mature here, I'd say, and they understand. Although they haven't been formally trained in the process . . . some of us have, some haven't. . . . The general trend is that people understand that you only block for highly principled reasons and . . . you are flexible, and you always look for the third way. All those things that make consensus work. People seem to have a handle on that here. . . . I'm pretty impressed with [our] collective ability to come to solutions.

Issues are not always perfectly resolved in these meetings. In interviews, individuals were split in their assessment of the villagers' decision-making process. Some believed the process works out well, while others felt that their village-mates did not always adhere to group solutions. Hannah, a young woman who lived on the property for over a year, voiced her frustration with the decision-making process:

> It seems to me that there are a few people that are really interested and involved half the time and most of the time they're the ones that take initiative, so they're the ones that end up making the decisions. Often there is a group effort made and there's an effort made to communicate so if anybody has something to say then they're welcome to say it. But, there have been many occasions where people felt passed up and wonder how things happened when they didn't know about it. . . . So, we're learning. It's all a process. And with people changing constantly that doesn't help the process grow.

As with all decision-making models, there are some potential problems with consensus. There may be hidden power relationships that cause some to defer to others; some may follow the group because they fear retaliation or other consequences; or some may go along with the group simply because it's the quickest and easiest option. It's worth noting that Emily, a property owner, found the process to work well, while Hannah, a newer resident, had some reservations. In this situation, there is clearly a power imbalance. To attend to these problems, groups using consensus decision-making should name and make explicit power imbalances within the group, especially as they relate to ownership, class, race, gender, age, or otherwise.

2.4.1.2. Work Parties

Essential to cooperative labor is the coming together of free farmers and, through nonwage and nonhierarchical labor, collectively maintaining the land. Ecovillagers hold periodic weekend work parties where community members clean and beautify the community. They work for hours taking care of the village, including chopping wood, collecting fruit, discarding unsightly debris, building things, and weeding. Afterward, community members enjoy a large meal and talk together.

I participated in a work party that constructed the driveway from the street to the woodshop. The workday began early, and individuals were free to participate as little or as much as they wanted. By lunchtime, two community members had prepared a meal that consisted of a green salad picked from the garden, quinoa, and a curry. Everyone sat around a lawn table near the garden to talk and enjoy the meal together. After an hour of relaxing, people got back to work in order to complete the project.

2.4.1.3. Land Ownership

In urban ecovillages, a major hurdle in the path toward restitution concerns the tension between the use value and the exchange value of city land (Harvey 1982; Logan and Molotch 1987). More will be said about this structural obstacle below. Modern landownership, as a condition of exchange value, obstructs ecovillagers' abilities to be free farmers. Two property-owning parties live at the ecovillage: Jamie and her son Ralph bought the five parcels where the ecovillage sits as a business venture about thirty years ago. Although once aspiring entrepreneurs, they no longer want to be landlords and therefore encourage others to buy into the property. Emily more recently bought into the property. The owners pay a mortgage and must ask residents for rent, thus reproducing capitalist economic relations by exchanging money for land and paying the bank. Although most decisions regarding community matters are made by the community as a whole, this landowner/renter situation interferes with the villagers' vision of relating to each other in a nonhierarchical manner. When times get hard, the landowners, as participants in economic institutions, face decisions about whether or not to sell parts of the property. Ralph expressed his dissatisfaction with being a landlord:

> I don't know who should own this place. I don't like being a landlord. I would like to sell off a portion of the property to get rid of my debt so I can just write. Ideally, I would love to sell it to the people in the triplex, but they don't have any money. I'm trying to find cool people who will buy into it. The rent from tenants almost pays the mortgage, taxes, and insurance, but I cover the rest in the form of credit cards.

Paying rent is an obstacle for the other villagers as well. Many ecovillagers must work in the city to pay for the land that they live on. Seventeen respondents spend some amount of time away during the day obtaining money for living expenses. If paying rent were not a necessity, villagers could devote more time to maintaining the garden, tending to the geese, or conversing with neighbors. While most of the working ecovillagers have jobs that do not contribute to pollution or land exploitation, such as working as a natural builders or permaculture teachers, others work for industries that do contribute to environmental degradation. One individual, for example, works on an assembly line to create large neon signs for other businesses.

2.4.1.4. Gardening

At the ecovillage, many residents care for and harvest their own portion of the collective garden. Produce from the garden is used for subsistence purposes and is often shared during community potlucks and gatherings. Because the purpose of the garden is subsistence, ecovillagers are committed to finding ways to keep it productive. Some strategies they use to maintain their garden include the creation of swales, or deep and narrow ditches between each garden plot that hold water well after the rainy season; rain catchment tubs for watering; and food scraps, compost, and chicken manure from their chicken coop to fertilize the soil. In addition, many ecovillagers utilize a technique for subsistence agriculture called permaculture, which I will presently describe.

2.4.2. Rational Agriculture

Ecovillagers subscribe to their own version of rational agriculture, or sustainable food production. Marx believed that the only way to restore metabolism between human beings and the earth is for cooperative laborers to create a rational agriculture. As defined by Liebig, rational agriculture applies the principle of restitution, "by giving back to the fields the conditions of their fertility" to "ensure the permanence" of the soil (Foster 2000, 153, 165, 169, 170). Ecovillagers follow a similar doctrine aimed at permanent agriculture called permaculture.

Permaculture, a term coined in the 1970s by Bill Mollison and David Holmgren, involves the development of "consciously designed landscapes which mimic the patterns and relationships found in nature, while yielding an abundance of food, fiber and energy for provision of local needs" (Holmgren 2004). In Holmgren's book, he emphasizes that "people, their buildings and the ways they organize themselves are central to permaculture. Thus, the permaculture vision of permanent agriculture has evolved to one of permanent culture" (xix). Holmgren identifies three key principles of permaculture: "care for the earth . . . care for the people . . . set limits to consumption and reproduction and redistribute surplus" (1).

A few teachers of permaculture lived on the ecovillage property when I visited. Of the twenty-four interviewees, eleven mentioned taking a permaculture class at some point. Emily, an ecovillage resident, taught permaculture at a rural ecovillage not too far from the city. When I asked about her political beliefs, Emily closely paraphrased Holmgren's principles:

> Permaculture. I'd call that somewhat of a political view, which is that we all need to become more sustainable where we are in order to protect the outlying areas. And, the foundation for permaculture is care for the earth, care for the people, and share the abundance. It's very simple.

Many ecovillagers have complex understandings of permaculture and sustainability that are reminiscent of psychosocial integration. But their views of connection are expanded to include environmental connections. They express the interrelated nature of each of the three principles written by Holmgren. Carol, a young mother and dome dweller, defines sustainability as encompassing the earth, personal relationships, and community:

> Sustainability is living in a way that enhances the quality of life for not just humans but for other species as well. So, a given area or land base can maintain health or increase in health over time. Biodiversity would increase for instance, or at least stay stable and not decrease. Sustainability in interpersonal relationships means that a relationship can continue, that when there's conflict there's a way to resolve the conflict. That goes for whole communities, that [when] there's conflict in the community, there's a way for the community to resolve that and continue on with each other, and people don't have to leave.

Sustainability in this view is multidimensional, with interrelated processes, including means of sustaining interpersonal relationships, community ties, and environmental integrity. From my observations and conversations with ecovillagers, I was able to identify actions that ecovillagers took to ensure the three permaculture principles in their everyday lives. Here I describe those in greater detail.

2.4.2.1. Caring for the Earth

Growing food for subsistence is one of the many ways ecovillagers attempt to care for the earth. Ecovillagers also express the importance of land stewardship, as in the case of a young man named Ears who worked in graphic design:

> I have been working the land a little bit at the office. You know, there's a little courtyard. I sometimes, at the beginning of the summer, I tinkered around with trying to grow different things at the office. My cucumbers didn't make it because it was too cool in the office. We have southern facing windows, but it didn't work out. So, even though I work in a pretty technological environment, I still try to keep that connection with the land. Especially there because, I mean that's where I spend most of my time. And I feel like it's honoring that piece of land to try to be, to live by my value of being a steward to the land no matter where I am.

Another way ecovillagers attempt to care for the earth is by avoiding excessive consumption. A critique of consumerism came out in about half of the interviews. Individuals distinguish themselves from other Americans by saying things like, "I'm not a consumer." A young woman resident explained the draw of the ecovillage for her:

> Probably the culture was like the final decision why I moved here. I just liked the people, and the mentality was a lot different than that of the Midwest, which was much more bourgeois in a lot of a ways. It's [the Midwest] very materialistic and middle class but contained in a certain box almost.

2.4.2.2. Caring for People

In order to sustain their cooperative labor, nourish community relationships, and maintain psychosocial integration, ecovillagers cared for each other in varying ways. These included holding regular community meetings, consensus decision-making, sharing food, having community potlucks, practicing NVC, and holding dispute resolution sessions. Ecovillagers also cared for people by trade. A few individuals specialized in what they called integrative intimacy, which involved getting at the root causes of individuals' emotional disturbances and finding ways to reintegrate their wounded parts. Further, some ecovillagers practiced a form of therapy and/or dispute resolution called co-counseling. In this type of therapy, each participant takes turns fully expressing their emotions while the other person listens and is supportive. During my regular follow-up visits to the community, I participated in an NVC reading group with three residents. As part of this group, we read Rosenberg's (2005) book on the subject and discussed how we might implement this communication style in our conversations with loved ones as well as with acquaintances.

A large minority of villagers expressed the importance of emotional wellness and expression. In her critique of American society, an integrative intimacy coach on the property expressed one of the problems she has with our culture:

> Not having feelings. You're not supposed to have feelings. You're not supposed to cry or even be ecstatically happy because it's upsetting to whoever's around. It would disturb someone or distract them from what they're thinking about or it might make them feel uncomfortable if you have big emotions.

In contrast with the larger culture, ecovillagers made a conscious and continuing effort to be open and supportive of one another's needs and feelings.

2.4.2.3. Sharing the Surplus

Sharing took many forms, and encapsulated Holmgren's third permaculture principle, sharing surplus. Community meetings were ceremonious potlucks that enticed villagers with community interaction and food. Another avenue for food sharing came from one community member who located a local bakery that gave away bread at the end of each day. He regularly brought extra bread back to the community to share. Ecovillagers also shared tools, knowledge, and other resources. During my stay, I found most individuals to be quite generous with their time, ideas, and collaborative work ethic. Nevertheless, some barriers existed to the sharing.

While permaculture advocates espouse caring for others and sharing the abundance, the permaculture courses that Emily helped arrange at the rural ecovillage cost a considerable amount of money. For almost a month's worth of courses, housing accommodations, and food, the courses were on sliding scale between $1800 and $2600. Some individuals could arrange some work-trade instead of paying the full amount, but for someone on a budget with debt or other expenses, taking a month off work and paying a thousand dollars for a few courses would be out of the question. The fact that nearly half of my interviewees were able to attend these courses speaks to their level of affluence. Consequently, teaching these courses is how Emily received a paycheck, which she needed for daily expenses. Thus, in a market economy, permaculture as a practice may be inaccessible to poorer members of the community, ironically undermining the caring and sharing aspects of the permaculture message.

2.4.3. Eliminating the Antagonism between Town and Country

Foster identifies three things that need to happen in order to eliminate the antagonism between town and country (2000, 169, 175). First, labor cooperatives must disperse themselves more evenly between both urban and rural areas. Second, an integration of industry and agriculture must occur. Finally, cooperative laborers must restore the soil by recycling human and industrial waste for soil nutrition. Ecovillagers do these three things by bringing agriculture back to the city, engaging in cottage industries related to the land, and composting.

2.4.3.1. Even Dispersal of the Population

The population dispersal proposition is debatable. Some scholars argue that more densely populated areas are better for the environment because density allows for more resource sharing (Owen 2009). Rather than distributing the population across rural and urban spaces, communities can instead reintegrate aspects of the country into the town, as in urban agriculture. Considering this, the ecovillagers alone cannot contribute to a more even dispersal of the population between the city they live in and the surrounding rural areas. They are restricted by urban zoning laws, codes, and land rent. This, in fact, represents a structural challenge in their restitution process, as discussed below. However, they do attempt to change urban living arrangements in order to facilitate a more sustainable use of resources. The design of the ecovillage, cohousing, and community resource sharing allow ecovillagers to use fewer resources individually and maximize efficiency with the resources they do use without increasing total resource use.

The ecovillage is planned to foster efficient resource use and community. Housing is situated around the perimeter of the property with workspaces in the center. This design forces individuals to interact with each other during work, because gardening and building are done in close proximity and all community tools are located near central workstations. Many dwellers on the property use common areas. The five dome dwellers share a dome support house that has a living room, kitchen, and bathroom. Similarly, tent dwellers share a covered outdoor kitchen, bathroom, and living room near the community center. Houses on the property usually house between two and four people. These cohousing situations facilitate community while requiring villagers to share resources. High-density resource sharing is more efficient than single-occupant residencies because fewer household necessities sustain more people. For example, the five dome dwellers share one kitchen, which necessitates only one set of pots, pans, and other cooking items. If each dome dweller lived alone, they would each need their own set, for a total of five sets of kitchenware.

2.4.3.2. Integration of Industry and Agriculture

Ecovillagers attempt to transform city relationships beyond the immediate ecovillage. In this particular ecovillage, villagers attempt to connect with their neighbors by inviting them to workshops, potlucks, and community events. They grow and share food in the middle of a city. The village sits between a neighborhood to the east and industry to the west. Ecovillagers also create a microcosm of this living situation within the ecovillage where they live, grow food, and have a woodshop area for building purposes. I use the word industry in two senses: (1) general business activity and (2) energy devoted to a work task.

The business activities surrounding the ecovillage include, but are not limited to, an ice-cream factory, retail services, restaurant services, automobile repair, and manufacturing plants. Ecovillagers attempt to make their city more sustainable by changing the physical landscape of an urban neighborhood block. The block is consciously designed with dwellings around the perimeter and gardens in the center so that people can come together in the middle to work and socialize. Villagers bring subsistence gardening to the city, thus incorporating aspects of the country in their city. Gardening and building are both done on the property, creating a microcosm of the integration between agriculture and industry. Finally, villagers repurpose waste, like wood and metal scraps, that they scavenge from local industries and dumpsters to make their homes.

Industry also refers to any work individuals devote to a task. In this case, industry can mean building earthen dwellings on the ecovillage property, growing food, building solar fruit driers, and any number of activities to which ecovillagers devote their energy. They built many dwellings on the property, with the exception of one house that sits on the eastern side of the property. Ecovillagers build icosahedral huts, cob houses, and scrap wooden homes. Moreover, villagers devote many hours to their gardens, situated between their homes.

2.4.3.3. Restoring Soil Nutrients

Ecovillagers make use of their waste to ensure soil fertility. In particular, compost is central to this end. Ecovillagers compost food scraps, weeds, human urine, chicken manure from their coop, and wood chips from the woodshop. However, compost cannot be haphazardly thrown into a pile and left to rot. Individuals must care for compost by exposing it to the right amount of sun, allowing worms to work through it, mixing an adequate amount of food and yard waste, and turning the pile every so often so that different parts are exposed to the air. Residents throw their food scraps into five-gallon buckets and empty them into the compost when the buckets become full. Male visitors are also encouraged to urinate in the compost.

At the eco-fair, Ears participated in a humanure—that is, human manure—project. The project entails creating composting toilets with sawdust and worms that adequately rid human waste of toxins and turn it into a viable plant fertilizer. This project literally returns human waste to the soil as nutrients. A project like this needs careful management, as human waste is a potential biohazard. Thus, ecovillagers had yet to engage in this project at the village.

2.4.4. Challenges from the Political-Economic Opportunity Structure

In the above analysis, I outlined activities that urban ecovillagers do to achieve restitution. Transforming the town-country antithesis, however, is a project that urban ecovillagers cannot realize on their own. This was briefly addressed when discussing the issues of landownership and the even dispersal of population. Here, I further elaborate on challenges by introducing the concept of the *political-economic opportunity structure* (PEOS) (Pellow 2007; see also Clement 2011a; Ergas and Clement 2016) and drawing on critical urban scholarship.

Prominent urban scholars have discussed the maneuverability of grassroots movements in capitalist cities. In particular, growth machine theory

(Logan and Molotch 1987), which drew from Harvey's (1982) work on the tension between the use value and the exchange value of land in modern urban society, demonstrates how local political and economic elites constrain the patterns of land use in and around the city. In the pro-growth context of American cities, maintaining the use value of land (i.e., the ways land can satisfy basic human wants and needs) is secondary to the pursuit of exchange, or monetary, value achieved through land-use intensification projects (e.g., the construction of commercial centers). Meanwhile, even though priority is given to the exchange value of land, the tension between land's use value and exchange value is not uniform across time and space. As evident with the ecovillage case study, there are spaces in which environmental movements can pursue restitution and prioritize the use value of land. Nevertheless, there are still limitations to, and structural constraints imposed on, the process of restitution and the realization of radical sustainability. Thus, drawing from the work of critical urban scholars, I frame the possibilities for and limitations to restitution pursued by ecovillagers in terms of the PEOS of an urbanized capitalist society.

To elaborate on the context in which ecovillagers work toward restitution, I draw on the notion of the PEOS first discussed by Pellow (2007; see also Clement 2011a; cf. McClintock 2014; Ergas and Clement 2016). Pellow argues that the prevailing view in social movement literature (i.e., the political process model, or simply *political opportunity structure*) cannot adequately explain the diversity and status of environmental activism around the world today. As a concept, the political opportunity structure prioritizes the nation state as the main institution in which social movements use conventional political means to carry out their agenda for change. Instead, according to Pellow, an analysis of environmental activism must acknowledge the symbiotic relationship between political *and* economic forces that characterize capitalism and how these forces shape activism. Thus, the PEOS emphasizes that activists' responses to environmental change are shaped by the broader political-economic context. Stated differently, environmental activists, including urban ecovillagers, cannot simply seek out and create opportunities for change *independently* of the constraints imposed by the larger political-economic structure of an urbanized capitalist society.

To make connections between Pellow's concept and the notion of restitution in metabolic rift theory, I draw from the urban growth machine literature. This latter framework provides a conceptual foundation for understanding the political-economic structure in which ecovillagers work towards restitution, ultimately with an emphasis on the use value of land. There are two *interrelated* structural obstacles in the path toward restitution. On the one hand, the first obstacle has to do with the intimate connection between modern urban areas and capitalism (Anderson 1976; Harvey 1982). On the other hand, the second concerns the conflict between the use value and exchange value of land in a capitalist city (Logan and Molotch 1987). In the context of ecovillages, the analysis of these obstacles comes from an environmentally informed interpretation of the urban growth machine (Ergas and Clement 2016).

With respect to the first obstacle, Harvey (1982) argues that alienation from the means of production, displacement from the land, and population concentration in urban centers are interrelated characteristics of capitalism. Under capitalism, people are driven into urban areas, thereby being deprived of access to the land. Harvey maintains, in particular, that landownership and rent are two mechanisms that prevent self-sufficiency by preventing "labourers from going back to the land and so escaping from the clutches of capital" (381–82). This inability to go back to the land preserves peoples' dependency on businesses for meeting their basic needs. In addition, private property and rent discourage the dispersal of human populations beyond urban sprawl and the development of new farming communities. Anderson (1976) articulates why there is a limit to the dispersal of the population under capitalism, writing: "Decentralization, outside of urban sprawl, is not profitable" (190). As such, landownership and the development of private property should be seen as structural obstacles to the restitution of the metabolic rift and the actualization of radical sustainability (see also Clement 2011b; Ergas and Clement 2016).

According to the urban growth machine, land uses in a capitalist city are largely determined by the exchange value of urban land, that is, the land's commercial value. American cities in particular are growth machines, characterized by dense, high-intensity land uses that increase aggregate rents and create wealth for the elite (Logan and Molotch 1987, 50). Logan and Molotch do not explicitly acknowledge the consequence of this structural dimension for urban agriculture, which is generally unsuccessful as a result of it. Despite the use value of urban agriculture, food production in cities does not generate rents that competing land uses do. This tension is, at times, dramatically played out in American cities when community gardens have been bulldozed in the face of a relatively well-organized social movement (e.g., von Hassell 2002; see also Sbicca 2012, 2014). The documentary *The Garden* portrayed such a battle, in which a farm in South Central Los Angeles was lost despite widespread support and attention from celebrities and politicians like Daryl Hannah, Danny Glover, and Dennis Kucinich.

In the Asaṅga ecovillage, most residents struggle with other economic obstacles in addition to rent. While many of them obtain free food via personal gardens, food stamp programs, low-income community food boxes, dumpsters, and fruits and vegetables gleaned from neighborhood fruit trees and gardens, they still purchase some food from the grocery store and local farms' CSAs. The quinoa and curry served after the work party I attended reflect the fact that villagers still participate in global economic conditions, as these foods are likely imported to the United States from other parts of the world. Additionally, space constraints in the village limit the types and amounts of food that can be produced. Ecovillagers in turn must purchase some foods from grocery stores. While ecovillagers do not sell the produce from the village, they do trade with some local farms and ecovillages nearby to obtain milk, some meat, and chicken manure.

The notion of the PEOS also emphasizes the way participation in the ecovillage is influenced, at least in part, by class. Class dynamics in the United States may preclude the possibility for some individuals to live in the ecovillage to begin with, because many of them (though not all) benefit from class privilege and its accompanying cultural capital. A political-economic opportunity that many ecovillagers have is the privilege of coming from middle-class backgrounds, since 70 percent report middle-class upbringings. This has allowed them the time to pursue higher education—bachelor's and in some cases master's degrees. In addition, many of them have had the luxury of time to travel around the United States to other ecovillages to learn earthen building techniques and to take permaculture courses, which often cost quite a lot of money (even though some ecovillagers are only able to participate in these courses as a result of sliding-scale prices and work-trade). One notable exception I met was an elderly man who received disability benefits from the state and had grown up in foster care. He lived in a small room and paid nothing for rent except for what he traded in work around the village. The US class structure certainly limits participation in this ecovillage, even if they attempt to be more inclusive.

2.5. Putting It All Together

The ability of Asaṅga to achieve radical sustainability and pursue grassroots innovations is limited by the larger PEOS. Sites like the Asaṅga ecovillage challenge city and nature dichotomies that inform traditional urban development policies and incorporate more holistic versions of sustainable development (Čapek 2010). Urban ecovillages represent an attempt to mitigate what Marx called the antagonism between town and country, because they bring components of the "country" to the "town" (Marx as cited in Foster 1999). More specifically, urban ecovillagers grow their own food and use their own waste as fertilizer in the city, thus breaking down the town and country dichotomy within the limits imposed by the town. Ecovillagers live cooperatively, make decisions based on consensus, form work parties, and practice permaculture—a type of rational agriculture with three principles that implicitly address the different forms of alienation: within and between individuals and society and our relationship with nature. While ecovillagers alone cannot eliminate the town-country antagonism, they do manage to more efficiently organize the space that they occupy and integrate waste into agriculture and their own building needs. Although there are checks against the landowner situation on the property, the ecovillage property is under a mortgage owned by three people who live there. While most decisions are made by consensus, when it comes to monetary matters, property owners have the final say.

I find that ecovillagers have more success in mending human-nature relations within the confines of their own village, even though some laws limit their internal activities. They have less success, and experience more structural barriers, when they attempt to mend the rift outside of this space. For example, they are able to use their own waste as compost in their gardens to nourish the soil from which they grow their own food. Outside of the village, they may have some success in utilizing industrial waste or garnering some resources from dumpster-diving, like day-old bread or bits of cardboard and wood used for building materials. However, they are still confined by the laws and economic norms of the larger society. In particular, many businesses padlock their dumpsters to keep divers out. Further, laws such as urban zoning influence the village's internal structure. Specifically, zoning laws restrict land use by regulating how many adults may live in a particular space. Portions of the ecovillage sit on parcels zoned for single-resident land use. Zoning laws don't allow the use of gray water for gardening purposes as well. Ecovillagers also are still subject to eminent domain laws, meaning their land theoretically can be seized in the name of state development. Neighboring industrial pollutants can contaminate their soil. Residents are still subject to land rent, or a mortgage. Because their work does not operate on an accumulation-based model, they have difficulty accessing enough money to pay rent or a mortgage. The very land they live on is in constant threat of being taken by the bank. In addition, they are limited by what they can reasonably grow in the small space to which they have access. The ecovillage operates on a relatively holistic systems perspective of sustainability that does not mesh well with atomistic, piecemeal approaches to urban regulations that dissect land into private property.

Despite these constraints, ecovillagers are able to mitigate pollution from waste, grow much of their own food, and remain persistent in their goals. In this way, this case study of an urban ecovillage demonstrates not only the limitations imposed by the political-economic structure but also the differences and similarities between the various forms of urban-based environmental action, in particular the growing experience of urban agriculture as an alternative food network. McClintock (2014), for instance, describes urban agriculture from two contrasting perspectives: on the one hand, according to many scholars, urban agriculture represents practical spaces of resistance against, and a radical alternative to, the agro-industrial food system; on the other hand, the emergence of urban agriculture can be seen as a response to the void left by a weakened social safety net in the wake of neoliberalism. Many advocates of urban agriculture frame their work, either explicitly or implicitly, in terms of individualism and entrepreneurialism, virtues that are consistent with the broader neoliberal agenda. While the members of the urban ecovillage may present a more systematic critique of capitalist forces and frame their work communally, many ecovillagers engage in a prefigurative social movement aiming to achieve a personal and village-community level transformation. Small-scale transformations, though important, stand in contrast to movements engaging in the necessary work of attempting to reorganize larger political-economic structures. There are notable exceptions, however, such as the ecovillagers working to change urban codes to accommodate ecovillages.

Moreover, most ecovillagers in the Asaṅga community are white and middle class. These aspects of their identities afford villagers privileges that other movement actors may not have, such as access to permaculture courses, university education, and travel to different communities around the United

States. Their whiteness serves as a buffer when they violate urban regulations and legal officials choose to look the other way. These privileges also afford ecovillagers the choice of downward mobility on their own terms, potentially as part of prefigurative social movements. Because my data are limited to one small, local ecovillage and my results cannot be generalized to all ecovillages or prefigurative social movements, many questions remain.

Could a holistic sustainability framework, such as an ecovillage, scale up in a capitalist context? How could some ecovillage practices reach a scale significant enough to bring about ecological or social change? Future research on these questions should consider not only the process of restitution toward radical sustainability—including transformative and regenerative socioecological change—developed here, but also the ways in which the larger political-economic context influences the actual practices of restitution. I explore these concepts in more depth in the next chapter, where I consider Cuban urban agriculture in the communist context.

3
Urban Oasis
Socioecological Sustainability in Cuban Urban Agriculture

Cuba, confronted with ecological and economic collapse during the 1990s, managed to radically alter its food production system to meet human needs and restore degraded ecosystems. Today, Cubans are celebrated globally for their innovative urban agriculture, which produces the majority of vegetables consumed on the island nation, and their urban reforestation programs, which preserve biodiversity, water and soil quality, and coastal areas. Cuba's particular history and culture created the necessary conditions for them to move forward with this nationwide experiment in agroecology and urban agriculture. Once a niche innovation in Cuba, agroecological urban agriculture was scaled up at a national level after their dual crises. Of course, Cuba's cultural and political-economic context differ greatly from that of ecovillages in the United States. But their struggle and resiliency carry powerful lessons for a planet faced with potential ecological, as well as economic, catastrophe and reveal a model for how to scale up more sustainable lifestyles.

In the previous chapter, I reflected on the political and economic landscape of an urban ecovillage in the United States to consider how they negotiate a social structure that sets limits to sustainability. In this chapter, I turn attention to the social aspects of sustainability—such as civil society, democracy, equity, and well-being—as well as the ecological, such as regenerative farming, in order to delve into their dynamic interplay.[1] Cuba serves as one of the few examples of a country with high human development indicators—that is, UN measures of life expectancy, educational attainment, and gross national income (GNI) per capita (United Nations Development Programme [UNDP] 2016)—that remains at the margins of hegemonic neoliberal globalization. After relations with their primary trade partner, the Soviet Union, collapsed following the dissolution of the USSR in 1991, Cuba's economy was considerably less connected to global markets than are most nations. Because their culture and political economy prioritized human well-being, Cubans

Surviving Collapse. Christina Ergas, Oxford University Press. © Oxford University Press 2021.
DOI: 10.1093/oso/9780197544099.003.0004

were able to harness human resources in a time of economic and ecological crisis. They collaboratively pulled together resources, ways of knowing, and labor, to collectively rebuild local and autonomous food production. As a result, many organic agricultural cooperatives were established, and Cubans gained relative food self-sufficiency. They have found that working in partnership with the environment is the most effective means of establishing ecological and social well-being. Indeed, Cubans' ingenuity has inspired others around the globe to extend financial aid and other assistance, and to bring these insights back home.

This case exemplifies what relocalization of food production, supplemented by global trade, could look like if we moved away from rampant consumerism and closer to needs-based consumption. Cubans consume less than their "fair share" of planetary resources: according to the World Wildlife Fund, the biocapacity of the planet to sustain each person is under 1.7 gha (WWF 2014). However, Cuba also demonstrates that living more sustainably does not mean that one must live an unhealthy, unhappy, or unfulfilling life. In fact, they reveal the opposite. Cubans have high human development and are among some of the happiest people on the planet (New Economics Foundation [NEF] 2013; UNDP 2016). While Cuba continues to have many problems, its shift to organic urban agriculture illustrates the complexity of moving toward a more sustainable and self-sufficient system. With a political economy geared toward meeting its population's needs, Cuba exemplifies one top-down communist attempt at sustainability that rejects neoliberal capitalism's no-matter-the-cost, competitive accumulation. Even so, there are structural and cultural limitations to socioecological justice and sustainability within Cuba that I discuss below.

Although the Cuban revolution lifted many out of poverty and guaranteed access to basic necessities, some groups still experience culturally entrenched social inequities. Women, LGBTQ, and Black Cubans face stereotypes and discrimination, as do nonhuman animals. In this chapter, I briefly review sustainability literature, and offer ecofeminist and environmental justice frameworks that explain the dimensions of socioecological inequity to provide context for the cultural barriers in Cuba's living laboratory. In addition, I describe Cuba's history, political economy, and culture to illuminate where they are in the process of moving toward socioecological justice, or a radical sustainability.

The UN's *Agenda 21* (1992), an action plan for sustainable development, set social equity—especially gender, income, and future generations—as a priority. Indeed, the plan states that the effective implementation of *Agenda 21* requires that women must be actively involved in economic and political decision-making (UN 1992). Because of space constraints and limitations in my data, I focus solely on gender inequity as related to divisions of labor and democracy, to examine socioecological connections. A significant way in which humans relate to their environment is through their labor, and the type of labor—such as farming, mining, or cooking—establishes ways of knowing and understanding local environments. Through a gendered analysis of a case of urban agriculture, I demonstrate how the social aspects of sustainability facilitate and/or constrain ecological sustainability by analyzing daily practices. I provide rich details that illustrate how democracy, well-being, equity, and civil society participation affect local environments. In the sections that follow, I briefly recap social sustainability and gender and environment theories. Then I provide a history of Cuban agriculture that preceded current urban agricultural practices. Finally, I delve into my fieldwork observations and interviews to lay out the socioecological aspects of sustainability in Cuban urban agriculture.

3.1. Social Sustainability

Social sustainability, as discussed in chapter 1, is the least theorized among sustainability's "triple bottom line" of equity, environment, and economy. Environmental justice and ecofeminist scholars have something to offer this undertheorized concept. These frameworks posit that socioecological justice requires both transformative as well as regenerative[2] approaches to socioecological change. For the purposes of this chapter, I examine how the four aspects of social sustainability—human well-being, equity, democratic government, and democratic civil society—relate to Cuba's agricultural transformation (Magis and Shinn 2009).

Although Cuba is not a democracy by Western accounts, the Cuban government puts human needs before profit and growth, in that all citizens are guaranteed food, healthcare, schooling, jobs, housing, and childcare. With these priorities already in place, Cubans were able to adapt to their dual crises. We can learn from their national priorities as well as the specific case of urban agriculture. Building capacity, and investing in human and social resources, is a necessary first step toward building resilient communities. Attending to peoples' needs allows them the physical and mental room to tackle challenges at hand. However, inequity limits capacity building, and thus resilience, because the contributions of devalued community members go unacknowledged or are actively stifled. Inequity poses multiple barriers to sustainability including ignoring valuable environmental knowledge, disregarding warnings of unintended consequences, and dismissing important contributions toward resilience. In addition, inequity creates social tension, which feeds civil unrest and drives social change. For clarity and length purposes, I focus on the gendered dimension of socioecological equity in Cuba in this chapter to expose how inequity creates barriers to resilience. In the next section, I review the literature on gender and environment to ground the gendered phenomena I witnessed in Cuba.

3.1.1. Gender and Environment

As described in chapter 1, political economy theories of environmental sociology, such as metabolic rift, are useful in the endeavor to create a more holistic definition of sustainability because they acknowledge how cleavages in social and ecological metabolic cycles are interrelated. Foster, Clark, and York (2010) argue that "the ecological rift is, at bottom, the product of a social rift: the domination of human being by human being" (47). Inequity, in essence, includes funneling resources—often environmental materials and money but also human labor and time—from those lower in the hierarchy to those at the top. While metabolic rift theory highlights how the exploitation of peoples and the environment are essential characteristics of capitalism (Foster 2000), the trappings of these practices did not disappear with Soviet or Cuban communism. Indeed, the Soviet Union and Cuba both ravaged environmental resources as part of their economies (Colantonio & Potter 2006; Maal-Bared 2005; McNeill 2000; Stewart 1992) and maintained masculinist institutions (Lapidus 1978; Shayne 2004). However, the interrelated aspects of sustainability—economy, equity, and environment—may interact differently in socialist and communist states and, for these reasons, are worth exploring.

According to Waring (1999), gendered—and I add anthropocentric, heteronormative, and white supremacist—economic institutions similarly affect both women and the environment through a valuation system that does not account for, or acknowledge, women's unpaid labor or ecosystem services, but rather treats them as free raw materials and exploits both ad infinitum. Material and cultural conditions, instead of essential gender characteristics, pattern gendered divisions of labor that affect how men and women interact with and experience local environments. Cultural norms in most countries position women as caregivers as well as subsistence and domestic laborers. The historical disregard for both women's work and the environment stems from value systems that were codified in world economic organizations created and dominated by elite white men (Merchant 1980; Mies 1998; Acker 2006).

Socialist feminists have widely theorized about the patriarchal nature of capitalism and capitalist institutions (Barbara Ehrenreich 1976; Holmstrom 2003). In particular, proponents of social reproduction theory suggest that capitalism perpetuates gender inequality and exploits women's work by treating women's biological and social reproductive labor—biological reproduction, unpaid care work, and domestic labor—as natural resources. Equating reproductive labor to natural resources serves to discursively and politically legitimate the exploitation of both women and the environment. The form women's exploitation takes is the invisible, unpaid labor that they do in their homes or in their communities to reproduce their "productive" labor power, or the paid labor they do for employment. Unpaid labor subsidizes capitalist enterprises and is a substantial part of the economy, even if unacknowledged (Holmstrom 2003; Bhattacharya 2017). Preoccupied with "productive" activities as represented by commodified labor and market exchanges, Western economists marginalize the unpaid household and community work usually performed by women. In this same vein, most commodity chain analysts have treated market/production and household/reproduction as discrete and disconnected spheres. By following this essentially sexist approach, these scholars deny that capitalists benefit greatly from externalizing the reproduction and maintenance of the labor force costs to households and communities (Dunaway 2014, 68–69). Dunaway argues that the distinction between reproductive labor and productive labor is no longer useful. Not only does this distinction serve to obfuscate how reproductive labor subsidizes market economies, but reproductive labor is increasingly entering the formal economy as paid labor, albeit poorly paid (e.g., servants, nannies, and cooks), thus productive.

Similar to social reproduction theory, Salleh (2010) attempts to integrate the invisible and most marginalized class of laborers into analysis for a larger critique of capitalism. She emphasizes the regenerative reproduction work that subsidizes the capitalist economy, which she terms meta-industrial labor. This is the unpaid work of caregivers, peasants, and Indigenous gatherers that has regenerative, rift-healing properties, such as tending to soils and managing forests. Anthropocentric economic measures of value, like use value, or material utility, and exchange value, or market worth, do not account for a flourishing ecosystem that is the basis of life itself. Salleh argues that debt and unequal exchange are part of capitalism's social rift–generating properties (211). She describes capitalism's social debt to exploited workers who experience a social rift by giving their lives and labor to capitalist production. Capital also creates an embodied debt to women and mothers who lose their intergenerational livelihoods, which take the form of handing down tradition and safe, natural environments to their progeny. Further, capital is ecologically indebted to global peasants and Indigenous groups who have lost their land and livelihoods in the face of industrial development.

We know less about how heteropatriarchy, white supremacy, and speciesism—as systems of inequity—work within other political-economic contexts, such as in a self-proclaimed communist nation like Cuba. Colonialism and capitalism coevolved as systems premised on inequity and have proliferated these inequities around the globe. And while capitalism has coevolved with different forms of patriarchy to create unique gender hierarchies particular to nation and place, patriarchal institutions also exist independently of capitalism in other postcolonial states. Indeed, Latin American feminists have acknowledged, "sexism was not the outcome of capitalism and imperialism but rather was shaped by a relatively autonomous, patriarchal sex-gender system" (Sternbach et al. 1992, as cited in Smith & Padula 1996, 184). Thus, when it comes to gender inequality, capitalism is not the only perpetrator. Intersectionality scholars assert that no one form of oppression—colonialism, classism, heterosexism, racism, speciesism—is the most pervasive or fundamental, but are dependent on particularities of place; they are interlocking and mutually reinforcing (Young 1990; Crenshaw 1991; Collins 2000; Pellow 2014). Outside of capitalism, patriarchal institutions and sexist interactions persist to (re)produce cultural prescriptions of gender difference and inequity in relation to other economic systems as well.

Similar to Cuba, the early Soviet Union saw emancipatory gains for women, such as accessible childcare, reproductive rights, and job security. However, as Holmstrom (2003) notes:

> most of these gains were eliminated later on and women were certainly not emancipated in the Soviet Union. But this does not show, as many commentators would have it, that "socialism failed women." Men were not

> liberated either. This was far from the socialism-from-below that the classical Marxists had envisioned. (4)

While some socialist feminists maintain their faith in an egalitarian socialist order, I am not so easily convinced. Cuba did not follow the same trajectory as the Soviet Union, remains a communist state, or market/planned-economy hybrid, and endures distinct cultural and structural (both political and economic) expressions of gender difference. This difference, I argue, is culturally entrenched, and, just like all other forms of oppression, cannot be solved by eliminating another form of oppression, such as class domination or capitalism (Ergas 2013a, 2013b, 2014).

I build on these theories to offer an analysis of the reproductive rift in Cuba and the movements in place to change it. Reproductive labor subsidizes agricultural production, as care work, self-provisioning, and subsistence work must be done in order for individuals to be able to cultivate in the fields. This work includes preparing food, taking care of children and the elderly, cleaning the house, and maintaining familial physical and emotional well-being. While the Cuban government did much to mitigate material inequalities between men and women, the government did not eradicate the socially restrictive roles that position women as the primary domestic laborers and caretakers. These culturally restrictive roles, I argue, have consequences for women's involvement in local democratic decision-making processes, limit their agricultural work, and obscure their local knowledge and contributions. Ultimately, restrictive roles have consequences for sustainability projects more broadly. My ethnographic research highlights the ways in which women's work is essential for urban agriculture (Ergas 2013a, 2013b, 2014). In the following sections, I outline the political, economic, and cultural context of Cuba generally to situate a particular case of urban agriculture. I discuss the history of agriculture and gender relations to show how they have shaped socioecological sustainability at one urban farm.

3.2. Cuban Context and Socioecological Sustainability

3.2.1. Colonial History of Agriculture

Cuba's colonial history is the history of European expansion elsewhere. After Christopher Columbus landed on the island in 1492, the Spanish proceeded with the brutal genocide of Indigenous peoples and the theft of lands and resources. The Spanish fully conquered Cuba by the beginning of the sixteenth century, and, by the mid-sixteenth century, they had largely destroyed the island's Indigenous Taíno civilization. There is evidence that, before they were overtaken, the Taíno helped the Spanish cultivate tobacco plants and taught them how to make and smoke cigars. Tobacco was one of Cuba's primary cash crops. In time, the introduction of enslaved peoples made sugar processing and coffee growing possible, and these became important exports as well. During the nineteenth century, a slave economy allowed Cuba to become one of the most prominent world sugar producers before slavery was abolished at the end of the century.[3] The Spanish decolonized Cuba after the Spanish-American war in 1898, and the United States maintained military rule for the next few years. Cuba finally gained its independence in 1902. But US citizens quickly established ownership of much of Cuba's land and resources, until Fidel Castro's revolutionary victory in 1959. However, even after their transition to communism, sugar plantations remained an integral part of the Cuban economy until the 1990s (Rumbaut and Rumbaut 2009; BBC 2017).

Cuba became economically reliant on the Soviet Union after the revolution and the subsequent imposition of the US embargo, initiated by President Dwight D. Eisenhower in 1960. Before then, Cuba's economy had relied on high-yielding sugar exports, as a result of Spanish colonialism and over fifty years of US political and economic domination. Thus, after the revolution, Cubans negotiated the exchange of sugar exports for Soviet petroleum and currency and continued industrial sugar monocultures in the form of large state farms (Colantonio and Potter 2006; Koont 2011). When the Soviet Union collapsed in the early 1990s and the United States tightened trade restrictions through the Toricelli Act in 1992 and the Helms-Burton Act of 1996, Cuba lost 70–85 percent of its economic support and export markets (Maal-Bared 2006; Koont 2011). Cubans refer to the ensuing economic crisis and period of resource scarcity as the "Special Period in the Time of Peace"; this period included shortages of food, fuel, and medicine.

After the revolution, Fidel Castro nationalized production and expelled US businesses from the island nation. Unencumbered by US-style corporate influence, the Cuban government developed state-run industries and farms. While many industries made products for export and trade, the primary focus of trade was geared toward meeting Cubans' needs rather than fattening their bottom line. Most trade occurred between Cuba and the

Soviet Union, and the Soviet Union collapse largely isolated Cuba from the global economy. To make matters worse, US-imposed trade restrictions not only prevented trade between Cuba and the United States, but also inhibited the docking of cargo ships from other nations as well. Specifically, these restrictions dictated that if a ship docked at a Cuban port, then it could not travel to the United States for another six months. This was an effective deterrent to trade with Cuba for many countries, given that the United States has a much larger and wealthier consumer market.[4] The revolution and the US embargo together positioned Cuba at the margins of globalized neocolonial development, which subsequently hurt their failing infrastructure.

In contradistinction to the United States, Cuba faces perpetual resource scarcity, not only from limited national natural stores, but also as a result of the US blockade that restricts its access to trade in the global market. They materially lack the resources necessary to create a consumer society. Indeed, they often lack the resources to maintain basic necessities. While Cuba has found other trading partners, in particular obtaining petroleum from Venezuela and machinery from China, the US embargo has resulted in huge financial losses for the Cuban economy. Between 1989 and 1993, Cuba saw about a 35–50 percent decline in GDP[5] (Koont 2011). During the Special Period, Cubans were faced with food scarcity and malnutrition. In order to survive, they had to revamp their food production systems and were forced to make revolutionary transformations in industry and agriculture (Rosset 2000; Companioni et. al. 2002; Lewontin and Levins 2007; Koont 2009, 2011; Rosset et al. 2011).

3.2.2. Necessitating Sustainable Development in Cuba

As a legacy of Spanish colonialism, Cuba's economy relied on high-yielding industrial sugar monocultures for export until the Soviet Union collapsed. Large-scale monoculture is a modern industrial agriculture practice where one crop is produced in one field year after year. This type of agriculture generally depletes soil nutrition and thus requires artificial fertilizers and eventually leads to the loss of fertile topsoil (Shiva 2005), creating a metabolic rift in the soil-nutrient cycle (Ergas 2013ab, 2014). It also required Cubans to import agrochemical pesticides and herbicides to control pests and oil to run heavy machinery. In 1989, three times more arable land in Cuba was utilized to produce sugar for export than food for national consumption, and most

Cuban food was imported (Ergas 2013ab, 2014; Koont 2011). The combination of the Soviet collapse and US-imposed trade restrictions destroyed the Cuban economy almost overnight. Because Cuba lost most of its economic support and export markets, Cubans no longer had access to the petroleum required to maintain large-scale agriculture (Koont 2011; Maal-Bared 2006). To make matters worse, the end of trade with the Soviet Bloc resulted in food scarcity, which reduced Cubans' average daily caloric intake and decreased their protein intake by 30 percent (Rosset 2000). Facing potential starvation, Cubans launched a revolutionary agricultural transformation (Koont 2011; Lewontin & Levins 2007; Rosset 2000).

Cuba's agricultural transformation began a process of socioecological regeneration (or restitution as in the previous chapter), whereby they attempted to reintegrate waste as nourishment in food production (Clausen 2009). Such projects included experimenting with agroecology, a technique that takes advantage of ecological synergisms using biodiversity, such as integrating nonhuman animals into rotational grazing systems with crops, diversifying with polycultures, and employing beneficial insects for biological pest control. Cubans also began recycling sugarcane waste as cattle feed and combined cow manure with worm castings to apply to soil as fertilizer, thereby restoring ecological interdependence and minimizing the need to import fertilizer from abroad (Clausen 2009; Koont 2011). Their experimentation also included creating urban organopónicos—growing plants in raised beds of organic materials confined in rectangular cement blocks in areas with poor soil quality. Additionally, personal household plots, which had long existed but gone unacknowledged within urban areas, took on renewed importance (Clausen 2009; Premat 2005). Altogether these experiments and projects served as the foundation for greater self-sufficiency, a system of urban agriculture, and a more sustainable form of food production (Ergas 2013ab, 2014).

Whether or not as a result of need, Cuba would not have been able to make these changes if it were beholden to a neoliberal global market that mandated structural adjustment programs and imposed austerity. Moreover, the outcomes of this crisis could have been catastrophic. Rather than food riots and revolutions, like those experienced in Egypt, Haiti, and Somalia during the 2007–2008 world food crisis (Shiva 2016); or tightening their authoritarian grip during mass deaths due to starvation, such as in North Korea after the Soviet collapse (Pfeiffer 2006); or imposing austerity measures, Cuba maintained its commitment to meeting people's basic needs by rationing limited resources. Cubans harnessed their human resources and ingenuity to collectively overcome this crisis.

While there is debate on whether or not Cuba imports the majority of its calories and protein (Altieri and Funes-Monzote 2012; Reuters 2012; World Food Programme 2017), urban agriculture has increased food security and sovereignty in the areas of fruit and vegetable production. According to Reuters (2012), in 2005 Cuba was "importing 60 percent to 70 percent of what it consumes [mostly so-called bulk foods] at an estimated cost of $1.5 billion to $2 billion annually." However, urban agriculture within and around Havana accounts for 60–90 percent of the produce consumed in the city and utilizes about 87,000 acres of land. Urban agriculture in other provinces supplies up to 100 percent of local produce consumed in surrounding communities (Companioni et. al. 2002; Premat 2005; Koont 2009, 2011; Raby 2009).

The Cuban National Group for Urban Agriculture defines urban agriculture as the production of food within the urban and peri-urban perimeter, using labor-intensive methods that pay attention to the human-crop-animal-environment interrelationships, while taking advantage of the urban infrastructure with its stable labor force (Koont 2011). Urban agriculture is the production of food and fuel where the highest concentrations of people are, reintegrating town and country, and contributes more than just nutrition to the surrounding population. If properly regulated, it can contribute to the environmental health of the city as well (Premat 2005; UNDP 1996). Cubans employ various forms of urban agriculture, including gardens, reforestation projects, and small-scale livestock operations.

Cuba's environmental protection and agricultural innovations have gained considerable global recognition (Clausen 2009; Premat 2005). The WWF's Sustainability Index Report (2006), which combines the UN Human Development Index and Ecological Footprint measures (or natural resource use per capita), determined that Cuba was the only nation in the world that was living sustainably (World Wildlife Fund 2006). As a highly urbanized population, at 75 percent (Oficina Nacional de Estadisticas [ONEI] 2011), urban food production is the most practical and efficient means to supply the population with food.

During the Special Period, Cubans' caloric intake decreased to approximately 1,863 calories a day. In the midst of this food scarcity, Cuba ramped up food production and, between 1994 and 2006, increased urban output a thousandfold, with an average annual growth rate of 78 percent (Ergas

2013ab, 2014; Koont 2011). Specifically, Cubans cultivated 18,591 hectares of urban land in 2001; by 2006, this had increased to 52,389 hectares. As a result of these efforts, in 2005 the population's caloric intake increased to an average of 3,356 calories a day. The Special Period also saw a sharp increase in unemployment. The creation of extensive urban agricultural programs, which included centers of information and education, provided new jobs that subsumed 7 percent of the workforce and provided good wages (Koont 2009, 2011). In 2007, urban agriculture, almost all of which is organic, comprised approximately 14.6 percent of Cuban agriculture (Koont 2011). There are about 300 urban gardens in Havana, with 10,000 hectares used for cultivating crops. Cuba's urban agriculture has increased food production, employment, environmental recovery and protection, and community building. Now I turn to a discussion of a specific site of urban agriculture, in order to illuminate what socioecological processes facilitated sustainable solutions and/or created barriers to them.

3.3. Gaining Access

I made two trips to Havana, once in June of 2010 and again from December to February of 2011. My field research involved interviewing, working with, and observing scientists and farm workers at one site, el Organopónico. I read formal and informal organization documents and saw documentaries put together by Cuban feminist organizations on Cuban women's work in agriculture. I also had the opportunity to stay with a woman who worked on the farm.

During my first trip in the summer of 2010, I established contacts through Global Exchange, based in the United States, and Cuban organizations such as the National Urban Agriculture Group, the Federation of Cuban Women, and the Cuban Association of Agriculture and Forestry Technicians. I met an urban farm president during this trip at el Organopónico, and he invited me to come back for more research, which I did in the winter of 2011. In my final round of data collection, I worked on the Cuban urban farm, el Organopónico. While there, I conducted many informal interviews with the men I worked with, and fifteen semi-structured interviews, with thirteen women and two men.

While the farm president was enthusiastic about my participation at the farm, as was his daughter, who housed me, some other workers were skeptical. However, most Cubans I encountered were friendly and eager to talk to me. This may have had to do with the fact that the president seemed happy to have me there and asked that people work with me. Some Cubans seemed particularly interested in talking to a foreigner, especially a US citizen who was interested in their work. The young people I encountered were enamored with American culture. Some disclosed that they longed to live in the United States. When I had an opportunity to divulge my Cuban ancestry to my coworkers, many of them would get excited and claim me as Cuban. However, some people seemed to have no interest in talking to me, either because I was a foreigner or otherwise.

The atmosphere on the farm was very casual. The president often in jest told other workers that I was his niece. When I tried to refer to him in Spanish using the formal "you," he insisted that he was not above anyone nor was he that old, and thus I should use the informal "you." People on the farm referred to each other in informal ways and often called each other comrade, a gender-neutral and nonhierarchical form of address, likely a remnant from their ties to the Soviet Union. Working on the farm was generally a delightful experience. The temperature was usually between 70 to 80 degrees Fahrenheit; it was almost always sunny; and most of my work could be completed outside in a cool breeze under the shade. Below are some observations from the field.

3.4. El Organopónico: A Cuban Urban Oasis

El Organopónico, the farm I worked at, sits on eleven hectares (roughly twenty-seven acres) of land on the eastern periphery of Havana and is less than a mile away from the northern coast. It is nestled in an expansive neighborhood surrounded by high-density housing, with three- to four-story apartment buildings filling each city block. When you approach the farm from the street, there is a row of vendors selling products from the farm, such as produce, pickled vegetable packets, herb packets, and sugarcane juice. If you walk past the vendors into the farm, you immediately come upon the ornamental plant sales area. Here they sell art, decorative plants, and knickknacks.

The farm has many different stations, and each works in relation to another. Stations include a plant nursery, rabbit cages, vermiculture, livestock (chickens and oxen), and fields of vegetables. It also has an area dedicated to processing produce into items like tomato sauce, herb packets, and chopped garlic for the street vendors. The farm feeds the livestock; the livestock provides manure for the worms; the worms produce humus for the seedlings; the seedlings eventually go out into the fields; the final products are sold to the surrounding community; and the economic office accounts for the sales, reinvests in production, and distributes profits to workers. Thus, the farm and the workers are able to reproduce their labor. Each station fulfills an important ecological, social, and economic function.

Immediately to the left of the ornamental plant station is the covered outdoor meeting and lunch area. This is where people congregate to discuss farm business, eat, and take a break. The roof is made of thatch, or layered dried leaves and branches, and under it are rows of tables and chairs. Across a large path that people walk and drive through is the office area that houses three separate rooms for the office staff. The buildings are made from clay with corrugated metal roofs. The kitchen, next door, is built similarly and is attached to the repair shop, where machinists solder together parts of broken-down machinery to make tools for the farm. To the right of the metal shop, and across another large path, lies the area for seedlings, plant starts, and the plant nurseries that house them. Tropical shrubs and trees, like palm and neem, shade the front area and are scattered throughout the farm.

If you walk further back into the farm, you will encounter fields of vegetables, livestock, and vermiculture stations. The farm grows many types of produce, including several varieties of tomatoes, lettuce, eggplant, onions, corn, and garlic. They also grow some tropical fruits, such as mango and guava, and different medicinal plants as well as herbs like mint, basil, and chamomile. The soil is a dry, solid red clay that coats your clothes and shoes in a layer that looks like rust. It is not the most fertile soil, so the farm workers must enhance it with a rich mixture for their small plants containing compost, manure, rice shells, and worm humus.

3.4.1. Environmental Sustainability and Agroecology

Agroecology is the primary farming method utilized at el Organopónico. This approach to farming seeks to more passively facilitate, rather than actively manipulate, ecological dynamics. Agroecologists scientifically research and build upon traditional and Indigenous agricultural practices.

Miguel Altieri, a prominent professor and researcher of agroecology, has worked collaboratively with Cuban urban farmers to research and cultivate agroecology in the field. He defines the science of agroecology as "the application of ecological concepts and principles to the design and management of sustainable agricultural ecosystems" (Altieri 2009, 2). This approach attempts to promote above-ground ecosystem health as well as soil microorganisms to maintain biodiversity, as a means of stabilizing crop pest populations, such as diseases, fungi, insects, nematodes, and weeds. On his website, Altieri (2017) further explains the goal of agroecology, which is to provide "the basic ecological principles for how to study, design, and manage sustainable agroecosystems that are both productive and natural resource conserving, and that are also culturally-sensitive, socially just and economically viable."

As Altieri writes, the Green Revolution—which includes the use of petrochemical fertilizers and pesticides as well as heavy machinery—has been disastrous for both ecosystems and human health alike. In tropical environments, the effects of climate change are expected to severely affect weather patterns, including long droughts, heavy floods, and extreme heat. Some common agroecology techniques have proven to protect against severe weather events by maintaining soil integrity, biodiversity, and moisture while minimizing soil erosion. These practices include raised fields, terraces, polycultures, and agroforestry. In addition, agroecology has been found in many cases to be as productive as conventional or industrial practices (Altieri 2009). It is a means of maintaining local food self-sufficiency year-round in places with impoverished farmers and vulnerable environments. In addition, it is a regenerative form of agriculture that departs from industrial models in that it minimizes the amount of manipulation necessary to grow food in varied, and at times hostile, environments. While it is still a managed system, it is far less coercive than the domination-oriented industrial paradigm that forces growth by dumping synthetic chemicals on croplands.

Agroecology does, however, require a great deal of attention and care, and therefore labor. Even though there is great need for farm labor in Cuba, some major challenges exist to recruiting younger people, as well as women generally, to become farmers. Farming requires manual labor. Workers must dig into soil as well as move plants and dirt, which gets their hands and clothes dirty. However, gender expectations for women dictate that they maintain cleanliness and not engage in manual labor. These gender norms position farming as an undesirable form of employment for women, even though it pays well. Aside from gender norms, expectations in the home and with family keep women's employment levels relatively low.

3.4.1.1. Listening to Nature

Despite these cultural constraints, a woman, Maricela,[6] occupies a high prestige job at the urban farm as the scientist who manages insect populations. This job requires training and expertise in identifying beneficial and harmful insect, plant, and nematode populations as well as experimenting with bacteria cultures. She manages insect populations by introducing natural predators and by making and applying organic pesticides. She reminds me that she does not consider any insect a pest, because they all serve a function in nature. Some insects are merely more beneficial for the plants than others, and she cultivates beneficial ones, such as ladybugs, to release them at times when there are infestations of white flies or aphids destroying the crops, for example. She also experiments with other forms of pest control, such as push-pull systems, intercropping companion plants that repel or attract pests away from crop plants. Specifically, intercropping plants, like marigold and sunflowers, attract or repel pests, minimizing food-crop infestations.

Maricela walks around the farm daily and examines the soil and many of the plant leaves and stems. She does this to observe plant diseases, insects, and weed populations and to collect plants and bugs to feed the insects she keeps in her lab. This attention to agricultural biodiversity is a significant component of agroecology. It is also a necessary aspect of environmental sustainability, because it minimizes disease and degradation. It is a way of "hearing nature" and tending to its needs in a place-oriented manner. Greta Gaard (2017) promotes a "trans-species listening theory." This theory takes lessons from intersectional research that highlights how individuals with dominant identities tend to speak while those with subordinate identities are expected to listen. This research demonstrates that "speaking is associated with power, knowledge, and dominance" (2017, xvii). However, ignoring dying species, temperature rises, and degraded soils is how we continue to propel climate change, and perhaps learning to pay closer attention, or "listening" more, to our ecosystems is necessary for continued survival. As the farm grows, more attentive eyes and ears are needed to ensure the farm's health and well-being. I explore the need for more workers, and other aspects of social sustainability, in more depth below.

3.4.2. Social Sustainability in Cuba

The Cuban government has taken the task of sustainable development seriously, instituting many laws on social equity and environmental protection during Fidel Castro's years as president, including laws on gender equity. However, many challenges remain for Cuba, because its culture and individuals within it must unlearn the hierarchal dogmas of colonialism that situate some groups of people over others and over the environment (the same is true for many nations, the United States included). In this section, I expand upon the four overlapping aspects of social sustainability—human well-being, democratic government, democratic civil society, and equity—explored in the first chapter, to detail how social sustainability works in Cuba and at the farm I observed. For ease of analysis, I combine democratic government and civil society below.

3.4.2.1. Well-Being

Cuba has maintained a commitment to meeting people's basic needs through a number of social programs that include free food rations, healthcare, childcare, and education. As a result, Cuba ranks among nations in high human development, which includes life expectancy (at 79.9 it is higher than in the United States, which is 79.5), years of education, and gross national income (GNI) per capita (UN 2018). The food rationing system was created to curb food hoarding by those with the most money or resources. The system is based on the UN minimum calories per person per month. Before the crisis of the early 1990s, each person received a full four weeks of rations. However, because food rations were sustained primarily by food imports, rations were cut back during and after the Special Period. Employers stepped in to fill some of the gap, and jobs supplied subsidized or free meals to ensure a minimum of calories (Morgan 2006). This is true at the urban farm where they provide free breakfast and lunch to their employees.

Healthcare and education have remained free during and after the crisis, and Cubans were assured support by their government even when food was scarce. Education complexes became important centers of planning for an agricultural transition and training the new urban agricultural workers. The Cuban government's investments in its people paid off, because Cubans had the skills and ingenuity to restore food production and the economy after the Special Period. Although today substantive needs are met, some Cubans' needs for self-actualization are not being fulfilled. This is causing many enterprising young Cubans to leave the country in search of more fulfilling employment opportunities and better pay. The Cuban economy still suffers greatly from the US embargo and trade restrictions. Cuba depends largely on the growing tourism industry that brings in higher-valued currency to sustain social programs and rebuild failing infrastructure.

3.4.2.2. Democratic Government and Civil Society

The extent to which Cuba may be said to be a democracy depends on your definition. By Western standards, Cuba is a nondemocratic nation led by an authoritarian dictator under a communist regime. The 2017 Freedom House report, based in the United States, deemed Cuba "not free" in political rights or civil liberties. Cuba's government is considered an authoritarian regime in the 2012 Democracy Index, a UK-based index, and scores poorly on all its indicators—electoral process and pluralism, functioning government, political participation, political culture, and civil liberties (Economist Intelligence Unit 2012). Cuba ranks poorly for press freedoms, at 171st out of 179 countries, particularly for arresting and detaining dissenters (Reporters Without Borders 2012). On the Economic Freedom Index (2013), Cuba ranks as "repressed" on indicators such as rule of law, regulatory efficiency, limited government, and open markets.[7] Finally, according to the World Bank Worldwide Governance Indicators in 2012, Cuba scored poorly on government effectiveness, political stability and absence of violence, regulatory quality, rule of law, and voice and accountability. The Cuban government did restrict media—the internet in particular during my stay (though this is changing). It also goes through phases of cracking down on oppositional organizations (Luciak 2007). Yet, an overwhelming 90 percent of Cubans believe that their political system is more democratic than other nations, even if not by Western standards (Luciak 2007).

The Cuban government has established access to the resources necessary for an informed and active civil society. Rather than focusing on electoral politics, Cuba engages in substantive democracy by giving citizens more equitable access to social, cultural, and physical resources than found in more traditionally democratic nations (Luciak 2007). Education in Cuba is free to everyone at all levels, including higher education (Koont 2011). Cuba ensures people's basic needs are met with the food rationing system and free access to quality healthcare. Cubans are able to enjoy cultural events like ballets and plays for free or an affordable price. They also can directly participate in local levels of governance by nominating neighbors and directly voting for candidates. Cubans point out that in the United States, democracy is often stifled for those who are poor and own no property. Cubans' access to government positions is less limited by their ability to finance a campaign than it is for US citizens. Perhaps as a result of US-led opposition movements, or at least using opposition movements as a convenient excuse, the Cuban government directly silences those who organize against the Communist Party (Luciak 2007).

3.4.2.3. Centralized Decentralization Approach

Although democracy is arguably crucial for sustainable development, Cuba's sustainable agriculture initiatives were largely generated by a minority scientific and technocratic community. While they sought to incorporate insights from peasant farmers, and many agroecological practices are garnered from traditional farming practices, some technologies evolved from a larger global agroecology movement. Prior to the Special Period, scientists developing organic technologies found it challenging to integrate them on a large scale throughout Cuban agriculture. Farmers and most scientists were proponents of conventional, petroleum-intensive farming methods (Stricker 2007, 2010; Koont 2011). However, the collapse of the Soviet Union, Cuba's primary source of petroleum, forced Cubans to make revolutionary changes to their agricultural production system. While many scientists and farmers thought the shift to organic production would be temporary, lasting until Cubans regained access to petrochemical inputs, agroecology scientists remained hopeful that time would show farmers and scientists that organic technologies would be more fruitful and sustainable methods of farming (Stricker 2007; Koont 2011).

Through a process of "centralized decentralization" (Koont 2011, 53), the Cuban government sought to make organic farming methods accessible to the population at large, democratizing organic food production. They did this by breaking up large state farms into smaller, cooperative farmer-owned units and subsidizing urban land for resident farming. They also created government organizations, like the Ministry of Agriculture and the National Group for Urban Agriculture, to offer technical help, facilitate farmers' networks, and link farmers with research, education, and service centers. The Cuban government developed nongovernmental organizations (NGOs), such as the Association of Agricultural and Forestry Technicians and *tiendas consultados agropecuarios* (agricultural consultation stores; Stricker 2007, 42), staffed with agronomists dedicated to dispensing free advice to citizens engaged in backyard farming (Stricker 2007; Koont 2011). The government created incentive programs that allowed farm workers to take home profits they earned from selling their produce. They also expanded agronomy education at the university level (Koont 2011), and created educational campaigns targeting youth to generate favorable perceptions of farm labor (Stricker 2007).

The government made these changes out of necessity, and certainly these programs began as top-down and technocratic. However, citizen participation and management were crucial to growing and maintaining the projects that began to flourish. Moore (2007) argues that for sustainability projects to be successful in the long run, experts must "depend on citizens to define them" and "rational deliberation among citizens" is necessary to create and sustain such projects (228). Moore begins to address the social aspects of sustainability that are crucial to any sustainable system. The farmers at el Organopónico found unused land, took the resources and information offered, and made the project their own by including crafts, flowers, and care.

Another important aspect of social sustainability is gender equity. In Cuba, women are overburdened with work obligations and the second shift of reproductive labor at home and have less time available to commit to local democratic work or employment. In the next section, I explore gender relations in Cuba.

3.4.2.4. Social Equity: Gender and Democracy

Equity is an important component of social sustainability, because equitable distribution of resource and empowerment ensures the well-being of all community members and is highly correlated with environmental well-being (Agyeman, Bullard, and Evans 2003). Cuba was one of the 189 nations represented at the 1995 United Nations Fourth World Conference on Women held in Beijing that unanimously agreed that gender equality is a matter of human rights and therefore requisite for social justice (Momsen 2010). Indeed, Cuba was the first nation to sign, and the second to ratify, the Convention on the Elimination of All Forms of Discrimination Against Women (CEDAW) that came out of the Second World Conference on Women in Copenhagen (UN 1980).[8] By contrast, the United States still refuses to ratify CEDAW.

However, Cubans still struggle with aspects of their colonial history that rank women, Cubans with darker skin tones, and LGBTQ individuals lower on socio-cultural hierarchies. Because of space constraints and limitations in my data, I only focus on gender inequity in the following sections.

3.5. Gender in Cuba

After the 1959 revolution, the new communist Cuban government instituted significant structural changes toward equality, especially gender equality. Such changes took the forms of laws that mandated gender equality and the adoption of a large-scale (ostensibly) nongovernmental organization. Fidel Castro created the Federation of Cuban Women (FMC) in 1960 with the explicit intent to incorporate women into the revolution. Cuban women saw great political gains that guaranteed meeting most of their material and practical needs. The FMC oversaw programs for women and enforced a series of new laws, which included the 1975 Family Code that mandated men share equally in household responsibilities. The 1976 constitution, which dictated that "women enjoy the same rights as men," institutionalized the right to employment, state protection of the family, motherhood, and marriage, paid maternity leave, and the prohibition of all forms of discrimination, among many other protections (Shayne 2004; Vallina & Pages 2000, 21). The FMC focused on practical needs, including access to free and safe healthcare, contraception, abortion, free and subsidized childcare, free education, divorce, and literacy campaigns.

Cuban women have achieved many advances. They make up 50 percent of the population, are about 37 percent of the paid labor force, and occupy about 34 percent of leadership roles in the workforce (ONEI 2011, 2015; Vallina & Pages 2000, 24). As of 2015, women also occupy 48.9 percent of national parliament positions, which far exceeds the 19.6 percent of women in the US Congress. According to the 2015 UN Gender Inequality Index (GII), defined as "a composite measure reflecting inequality in achievements between women and men in three dimensions: reproductive health, empowerment and the labor market," Cuba ranked 62nd among the most gender equal nations (the United States ranks 43rd as a comparison; UNDP 2016).[9] Cuba ranks very closely to the United States on many of these measures, which is interesting, given that Cuba is a poor nation compared to the quite affluent United States. This demonstrates that even in the face of great scarcity, the Cuban government attempts to divide social resources more evenly.

In fact, Cuba fares better than the average for Latin America across all gender equality indicators except labor force participation.

Despite these strides, structural and cultural barriers to equality persist in Cuba. In 1985, the Third Congress of the Communist Party of Cuba realized that women and Black people held disproportionately few leadership positions and established affirmative action-like policies to promote equality (Prieto & Ruiz 2010, 164). However, soon after this, the Soviet Union collapsed, and the economic crisis began. The gains women made toward equality after the revolution began to decline, including reduced access to positions of power and increased household workloads. Pages (2008) contends that "a number of experts from the FMC's Women's Studies Center indicate that women are the ones who have been hardest hit by the difficulties of daily life in the Special Period" (313). Cuban equalization projects, like building homes for people who lived in dilapidated housing and daycare centers for working mothers, came to an immediate halt.

Dominant definitions of gender are produced, reinforced, and reproduced within institutions through laws, policies, norms, and interactions, and other behaviors defined as deviant are sanctioned accordingly. As Kimmel (2000) notes, institutions "express a logic, a dynamic, that reproduces gender relations between women and men and the gender order of hierarchy and power" (95). Culture, on the other hand, weaves together our codified institutional practices and interpersonal interactions to make and remake the gender order. In this way, culture is a set of "tools" that bridge the gap between structure and agency (Swidler 1986; Carruyo 2008). In Cuba's case, their gender hierarchy never fully went away, and then regained prominence during a time of material resource scarcity. Culturally prescribed gender roles in Cuba became more pronounced during the Special Period and contributed to women's declining status and inability to democratically engage in decision-making. The economic collapse exacerbated structural problems, including the loss of public transportation, which further limited women's career options. Services intended to ease domestic burdens were also subject to cutbacks after the Special Period. For example, daycare center construction came to a halt, and resource shortages caused daycares to close down at times, making it difficult for mothers to go to work (Pages 2008).

Cuban socially constructed gender roles assign women the undue burdens of unpaid reproductive labor. When resources became scarce, many women felt obligated to give up their permanent or temporary jobs or retire early in order to stand in line for hours to buy household necessities, deal with shortages, improvise meals, and care for children, the elderly, and the disabled (Vigil 2008, 310). Consequently, Cuban women worked three times as many unpaid hours as men (UNDP 2010). At times, women had to choose to leave careers they loved to take other jobs in order to earn more valuable convertible currency. Women's double shifts and other culturally specific forms of sexism, including societal devaluation of women, prejudice, paternalism, and women's internalized self-limitations, restricted their ability to be involved in local leadership and even paid employment (Pages 2008; Vallina & Pages 2000). Moreover, the culture of paternalism and patriarchy is deeply entrenched in Cuban political and economic institutions.

Men make up the majority of large-scale urban farmers in Cuba, which is common in commercial "cash crop" models of agriculture worldwide (UNDP 1996, 66–68), whereas women tend to cultivate in small-scale, self-provisioning spaces around their homes (Premat 2005). Thus, from an institutional perspective, women's work and contribution toward socioecological resilience is overlooked, ignored, or made invisible. As I detail below, these gender disparities are far more than an interesting side note, because women are in fact integrally important to any successful sustainability project. Indeed, dismantling limiting gender stereotypes could serve to increase farm labor, since agriculture desperately needs more workers. In the following sections, I examine data, including interviews, field notes, and local Cuban research and documentaries, to describe socioecological processes in sustainable agriculture.

3.5.1. Farm Labor and Gender

Cuba needs agricultural workers to engage in agroecological organic farming. However, young people do not want to become farmers, and parents do not favor a farming life for their children. The urban farm president stated:

> What is the real crisis within organic agriculture? The human resources. . . . With . . . whom are we going to make agriculture in the future? Where are those people? Where are they, the labor force? In Cuba, all the careers in the university are free. People pay nothing, but the least amount of students, they go to study agronomy. . . . What do I mean by this? OK . . . the culture of the Cuban average man, if a son tells me he is going to study agriculture, "You are crazy, boy." Why would you study agriculture? You could study

> architecture, yes, psychology, yes. Computers yes. Biology, but not agriculture, no. It is not a policy or state problem that convinces people to not want it. People reject agriculture. It's a reality.

The farm president's message about the lack of farm workers is corroborated in local documentary films. As I discuss in more depth below, Iset, an urban farm worker, and her family accompanied me to a documentary film screening and art show in Havana. One documentary features a Cuban farm family whose son quits college to work on the family farm. His parents are distraught by his decision to leave college. His mother laments, "It really hurt me, and I felt really sorry about it, but I said, okay, in the end if that is his decision. I explained to him, 'Son, my future is already set. Yours– you're the one that has to try to create it. If you decide that this is what you're going to do, take a walk in your mother's shoes, because it's no secret to anyone that working in agriculture, you have to work very hard.'" Another farm family speaks of their children:

> Mother: I would like for them to go to school, because the life of a farmworker is not an easy one. I tell them, "look at your father. He's out in the sun working all day long."
>
> Father: There are many of us who have made the choice to work in the farm, but we have to recognize that it is a difficult job. If they have other possibilities, I'd say no; I would not want them to follow in my footsteps.

Gender expectations further complicate this problem, because women involved in agriculture experience great societal hurdles. Specifically, some husbands are reluctant to allow their wives to go to work to begin with, and when women do, some face discrimination at work. In one documentary, a cowgirl describes the dual troubles she faced with work, first from her husband and then from her boss:

> I married a man who had three children . . . so, I took care of the three children. . . . He was a good father and a good husband, but I'm telling you that he had his macho issues. He didn't want me to work. . . . He would tell me that women who worked on the street were looking for another man on the street, so, well, we had some misunderstandings, some fights in the marriage, but I was able to overcome that, I was able to start working. . . .

> My boss didn't want me working there because I was too young. He would say that I didn't have the experience. I had a lot of children to raise, because he saw me as a . . . nuisance to the work, that I wouldn't be productive. . . . And I was a cowgirl, but I would never smell like cows. Do you know what I mean? I would always use perfume, I would dress up . . . and he would say that I wasn't the type of woman for these jobs. To the extreme that he would do things so that I would voluntarily leave, because he couldn't directly fire me.

However, this story ends triumphantly, as this cowgirl was able to prove herself to her boss in the end:

> The boss I had wasn't very friendly to me and he would always place me in hard places, maybe to see if I would respond or not, but I didn't respond until now. One day, he came up to me, because they came to ask him for an innovative worker and there weren't any male workers who were better than me. So, he had to give me National Vanguard standing. From that, he realized and he came and put his hand on my shoulder and told me: "You gave me a beating without even touching me. You have proven to me that you are stronger. You have proven to me that you are currently stronger than any man and that when women apply themselves, they can do whatever they want whenever they want."

Approximately 43 percent of working age women in Cuba participate in the labor force, in contrast to 69 percent of men (UNDP 2016). This disparity may be due to a variety of factors, including controlling spouses or families, excessive household expectations and reproductive labor, and cultural norms that routinely limit women's options. Changes in cultural perception around agriculture and gender could simultaneously assist urban agricultural growth and socioecological sustainability more broadly. However, even as many things change, gender divisions of labor persist.

3.5.2. Gender Divisions of Labor at el Organopónico

Gendered divisions of labor at the urban farm are clear in each sector. During my fieldwork, the total number of employees was 177, with about 40 women representing 23 percent of the workers. There were 13 managers, 4 of which were women who represented about 31 percent of the leadership positions. Table 1 breaks down each sector at the farm by gender.

As demonstrated in Table 1, more women than men worked in offices doing human relations, accounting, and commercial sales. Only men worked out in the fields tending livestock, composting, and harvesting. I also observed that women tended to cook, sweep, clean the grounds, planted seeds, watered small plants (though not in the field), packaged items for sale, and did administrative work. They were often sitting in the shade and/or inside a building, while men were doing physical labor in the hot sun. In fact, some men complained about how easy women's work seemed to them. Tasks typically reserved for men were engaging in manual labor in the fields, cultivating crops, operating heavy machinery, and working with large animals. In fact, I never saw a woman doing these types of activities. I am not suggesting that gendered divisions of labor are inherently problematic. However, it is clear that with these divisions of labor came certain cultural attitudes that limited women's options.

Cuban conceptions of gender affect the types of work that men and women do, and evidently contribute to the lack of women's participation in large-scale urban agriculture. Beliefs that men are physically stronger and that men are protectors of women limit what men allow women to do as well as what women choose to take on. Momsen (2010) affirms that "women are prevented

Table 3.1 Urban Farm Sector by Gender.

Sector	Women	Men
Economic and Human Relations	13	5
Service	7	1
Plant Start House	6	3
Commercial	5	5
Agro-industry	5	5
Ornamental Plants	4	8
Organic Material	0	6
Livestock	0	8
Watering	0	9
Maintenance	0	11
Security and Protection	0	26
New Farm Area	1	48

from entering certain types of employment, usually on the grounds of physical weakness" (178). During our interview, the farm president's daughter explained attitudes of farm labor. "All the farm men, those that work in the field, they are a little macho, and they think the fieldwork is hard, it's demanding, the women cannot do that. So, this slightly undervalues women." However, she argues that things are changing: "But nowadays there are thousands of examples of women achieving many things."

My personal interactions with coworkers illustrate the ways in which men asserted their perceived superior physical prowess. While at the greenhouse, I worked under a male-identified boss. When I performed "women's work," like planting seeds in trays, he mostly ignored me. If I tried to do "men's work"—shoveling dirt, handling wheelbarrows, carrying too many trays at a time—he made a point to tell me that I do not have to do this work, or that I can stop if I am tired. One day, we discussed my ability to perform these tasks. He said, "You can do the work that the men can do, but probably not as much because you're a woman. The men are stronger, but you happen to be a capable woman." As another example, I insisted on pushing wheelbarrows despite the resistance from my male-identified coworkers. Every time I picked up a wheelbarrow, for a total of five times, it caused a conversation. One day in particular, a male-coworker and I were transporting dirt with wheelbarrows. First, he told me that I didn't have to do it; then he insisted that I use the smaller wheelbarrow and not fill it up. As we walked together to transport the dirt, another older man saw that I was pushing a wheelbarrow while a man walked with me. This older man jokingly disparaged my wheelbarrow coworker by calling him a "capitalist exploiter" and added "I cannot believe you are making a 'girl' push a wheelbarrow."

3.5.3. The Second Shift

Tasks at the farm are clearly sex-segregated, yet the most disparate gender divisions of labor happen during the second shift. My field notes illuminate how women were affected by the second shift, or unpaid domestic labor done before and after paid work (Hochschild & Machung, 1990). Salleh (2010) argues that capitalist modes of production are subsidized by reproductive labor, generally assigned to women in most societies. The Cuba case illustrates that other economic structures can also rely heavily on the regenerative labor that women usually perform. Gender roles for Cuban women position them as the primary homemakers. I interviewed thirteen women at the farm, eight of whom had male partners. Six of the partnered women said that they did the majority of the housework. Some of the women I talked to accepted it as their lot in life to be the primary homemakers, even though they also worked all day at the farm. There were a few instances where women voiced resistance, but a defeated acquiescence was the more predominant disclosure. When asked how she and her husband distribute household labor, an accountant at the farm lamented, "Women have to carry more of the weight than men." Another young woman who spent most days preparing herb packets for sale uttered as a matter of fact, "I wash dishes every day because I am the only woman in my house." My findings corroborate research on time use in Cuba that suggest that women work as much as three times as many unpaid hours as men (UNDP 2010).

The term "uneven development" usually describes unequal power relationships between nations, specifically developed and developing nations (McGee, Ergas, and Clement 2018). However, it also can be applied to inequities within nations at the subnational level between men and women and between white Europeans and Indigenous as well as historically enslaved populations (Labao 2016; Tickamyer and Patel-Campillo 2016). For our purposes, uneven gender development specifically benefits men and other family members, who depend on interpersonal unpaid care work and reproductive labor performed disproportionately by women (Gaard 2017). According to ecofeminist scholars, women's unpaid labor is extracted and exploited at the expense of their personal well-being, with few resources for their personal care and reproduction. Neoclassical economics disguises the ways reproductive labor subsidizes households, labor productivity, and the national economy by neglecting to include this unpaid labor in national accounts (Waring 1999).

Because I had the privilege of staying with a Cuban family, I observed the distribution of domestic labor between a married couple, Iset and David. To be sure, both of them worked very hard. However, Iset did not have anywhere near the amount of leisure time that David enjoyed. In reference to the longer hours she worked, Iset revealed, "If I am reincarnated, I will come back a man." Here, I detail the amount of work that the Iset did compared to her husband.

The tasks delineated by gender divisions of labor disproportionately consumed nearly every moment of Iset's time. Her lack of time to herself illustrates the ways in which the second shift limits women's ability to participate in decision-making processes conducted during the farm workers' free time (Hochschild & Machung, 1990). Iset worked Monday through Saturday at the urban farm. Monday through Friday, she woke up at 7 a.m. to get her two sons out of bed and ready for school. She made them breakfast and dressed her younger son. After she had toast and instant coffee, she walked her younger son to school at 7:45 a.m. She had to be at work at 8 a.m., and she worked until 4 p.m. At 4 p.m., she left work to pick up her younger son from school. They usually stopped at a market on the way home to purchase items for dinner. Once home, Iset began preparing dinner. She first cleaned the beans, rice, and chicken she was going to cook. While doing this, she usually helped her sons with homework at the dinner table. Once she had the beans, rice, and chicken in the slow cookers, she began preparing lettuce and tomatoes for a salad. At 8 p.m., on days when her husband was working, she would bathe her younger son. Around 8:20 p.m., she would serve everyone dinner. Because the table was not big enough for everyone, she waited to eat until the others had finished, and meanwhile washed the dishes. Finally, she ate around 9 p.m. She would clean up after everyone and then watch a soap opera until she went to bed, usually around 11 p.m. On Saturdays, Iset worked a five-hour shift at the urban farm. Sundays she used to catch up on household chores like sweeping, mopping, and cleaning the bathroom.

During my stay, Iset was also finishing a university degree in agronomy. She was able to leave work for class some afternoons. However, if she had an exam, she often stayed up studying the whole night beforehand, because she had almost no extra time during the week for homework. The number of hours she worked combined with how little she slept and ate may likely have health consequences as she ages. Admittedly, most women I spoke with were not juggling work, childcare, and university classes. Iset is particularly industrious. Yet, even if she was not in school, she still would not have had time for much of anything else. Her complete lack of free time precluded her involvement in neighborhood organizations, like the Committees in Defense of the Revolution, civil society, or any other level of democratic participation.

David worked every other day at a restaurant as a cook. His shifts were typically twelve hours long, and he earned quite a lot by Cuban standards, because tourists, who paid with more valuable currency, frequented this restaurant. His typical day off consisted of waking up around 11 a.m. or noon. His mother-in-law would have breakfast ready. Some days he would sit and watch television or music videos for several hours. On other days, he would go to a garage on the urban farm and work on his car, which did not function the entire period I stayed with the couple. Around 8 p.m., he usually bathed his younger son. As soon as he finished the bath, his dinner was ready. After eating, he would watch television in his room.

During my visit, Iset averaged about 87 hours of work per week, and David averaged between 37 and 49 hours of work. While the instance of Iset and David is not generalizable, their experience corroborates other findings that suggest that women work longer hours during the second shift in Cuba (UNDP 2010). The work women do domestically actually subsidizes farm labor, because it allows many men to go into the paid workforce without consideration of household tasks and childcare. In addition, women's domestic labor disproportionately benefits men, granting them the time required for participation in various local democratic decision-making processes. Antithetical to social sustainability, the net effects of women's combined household obligations and lack of leisure time are the disproportionate exclusion of women from civil society; the potential health consequences of the stress associated with care work; and the devaluation of women's societal contributions. If women want to participate in democratic processes, they must give up any precious leisure time or lose more sleep. Culturally prescribed gender roles position women as primary domestic laborers and caretakers. These gender divisions of labor, though recognized in unenforceable laws like the Family Code, go unacknowledged in gendered work institutions and a patriarchal economy that neither remunerates nor documents the domestic or reproductive contributions of women. In addition, research on unpaid caregivers have found that caregivers help out family members in need at a considerable cost to their own physical and mental well-being (Moore, Zhu, and Clipp 2001; Pinquart & Sorensen 2003; Vitaliano, Zhang, and Scanlan 2003). Caregivers experiencing mental strain are at a higher risk of death than noncaregivers of the same age and sex (Schulz & Beach 1999). In light of these inequities, Cuban organizations are stepping up to offer gender sensitivity training in different communities around the island and are exhibiting their work in Havana.

3.5.4. Stories of Resistance and Change

As previously noted, inequity generally results in acts of resistance, and some Cubans are actively contesting these gender tropes. During my stay in Iset's home, she organized a family trip to an art opening and documentary screening in Havana. I was able to witness Cubans scrutinize gendered cultural beliefs at the opening with art and films that highlighted the gender divisions of labor in Cuban agriculture. The organization that supported the art opening, the Programme for Local Agrarian Innovation (PIAL), a subsidiary of the National Institute of Agricultural Sciences in Cuba, had a gender education strategy to change gender attitudes and promote gender equity. Under the patronage and guidance of the Swiss Agency for Development and Cooperation, the two organizations backed a group called Against Hegemony that put together the art show. They featured photographs of women doing agriculture work and films interviewing men and women farmers and their children. In one documentary, interviewers asked participants what they believed constituted women's work. The audience roared with laughter as the children rattled off a list of domestic chores. This was the only instance in my stay in Cuba where I saw Cubans critically discuss gender expectations collectively.

After the film screening, the Against Hegemony organizers generously shared a total of fourteen short documentaries with me, all about gender in Cuba and most about gender and agriculture. Several were filmed in 2007, and subsequent films explored cultural shifts in gender perceptions in different communities after PIAL programs were initiated. As a result of PIAL programs, some rural men admitted that they had taken their wives' work for granted in the past, and some rural women acknowledged that the program empowered them to become more vocal in their households and active in their communities.

One clip featured a female narrator describing a scene at an agricultural market where women are selling their canned fruits and vegetables, some for the first time. She comments on the changes in one rural community called Las Caobas, named after the mahogany tree:

> Moderator: The development of the family farms cared for by the women has changed the life of many in Las Caobas. . . . There isn't just a need to solve problems with instant solutions, but they also think about how to overcome critical situations or scarcity of some foods. That's why they store vegetables and fruits in these preserves which surprised both Jibareño men and women at the agricultural fair.

In another clip, a woman journalist documents changes in a different rural community where community members gather to learn about preserving food:

> Moderator: When I got there, I thought I would only find women, but reality was different. This morning, rural women and men from La Redonda Community met in the municipality of Urbano Noris to teach and learn about the interesting and currently vital world of preserves. . . . This time, we went beyond admiring the presentation and exquisiteness of the preserves. We did not know how much these rural women have changed their daily routines, which are no longer limited to taking care of the family and making sure that the man working in the field is well taken care of.

Food preservation is a crucial task, often performed by women on these farms. One of the films observed a community in the aftermath of a devastating hurricane. Homes and croplands were destroyed. What saved the rural community from starvation were canned foods and preserves prepared by rural women. A farm woman noted, "It's like food storage, as I say, like the ants, right? We're always storing and storing, because you never know what will happen tomorrow. There could be a hurricane. There could be something else, so that you have stored food and don't lack any." Rural women traditionally engaged in this task for their families, but men only began to acknowledge the importance of this work for food security when their wives began selling their wares for profit. One man spoke of his wife vending at a local market, "It's great to watch my wife sell here, because it's a good thing for the whole family."

Just as women on the urban farm suggest, these documentaries showcase the changes in cultural perceptions and expectations of gender over time. One film responds to an oppositional male interviewee who states "Men provide for women. What's always been doesn't need to change. And it's always been that way. Why would that change now?" The film text retorts, "Why would it change? Because it will change. . . ."

3.6. Conclusion

Socioecological sustainability and justice require that we move away from anthropocentric conceptions of sustainability to consider human and nonhuman relationships, connections, and exchanges. Metabolic rift theory contends that humans engage in material exchanges with the environment through their labor, and humans' labor provides a material connection to the land. Because humans' labor is a significant way in which humans interact with the environment, social inequalities that produce gender hierarchies and divisions of labor affect men and women's labor-mediated interactions with nature. Political economies and culture prescribe gender divisions of labor. As a result, men and women often interact with the environment differently; thus, their labor has different environmental consequences and, in return, their environments affect them differently. Salleh (2010) argues that women's caregiving reproductive labor has regenerative and rift-healing properties, while other forms of labor, such as mining and industrial farming, have rift-generating properties. Uneven development of paid and regenerative labor, whereby paid labor is more culturally and economically valued, can put undue burden or strain on lesser valued groups. Regenerative and reproductive labor is often performed for paid laborers at the expense of the caregiver. Undervalued groups usually push back in acts of resistance as a means of obtaining well-being. To develop socioecological justice toward a radical sustainability (that is at once regenerative and transformative), awareness is needed for human and nonhuman well-being, equity, and democracy.

Social sustainability literature advances four interrelated components, including well-being, equity, democratic government, and democratic civil society. Scholars identify democracy as an important feedback mechanism between communities and the environment. If there is environmental degradation, there are likely human and nonhuman communities at risk whose voices remain unheard. Attending to the needs of nature and the most vulnerable populations of people supports everyone's health and is mutually beneficial for human and nonhuman communities. Place-situated acknowledgement of nature is necessary to hear the needs of the environment. And equally valuing men and women's work is crucial because gendered labor divisions make men and women attuned to different aspects of environmental change or harm. Thus, equity, well-being, and democracy are advanced by listening to all stakeholders, human and nonhuman alike, and acting to resolve their problems (Norgaard and York 2005; Winslow 2005; Moore 2007; Boyce 2008; Ergas and York 2012; Pellow 2014; Gaard 2017).

Democracy in Cuba is complicated. Cuba is not considered a democratic or free nation by Western standards. Indeed, the Castro family ruled over the nation from the communist revolution in 1959 until 2018. However, the government ensures that all Cubans' material needs are met, for food and shelter, as well as education, employment, healthcare, and childcare. As a result, many Cubans are able to lead active and community-engaged lives. There are inhibiting factors that affect different groups of people. Specifically, women's domestic roles in Cuba place high demands on their time and labor and affect their ability to actively participate in employment as well as local democratic processes, or civil society. Further, women's contributions are routinely overlooked or made invisible in their economy. Notwithstanding, Cuban organizations are working on gender strategies to change these problems and ensure that women have greater access to and more active civic participation. Greater gender equity and democracy are mutually reinforcing and enhance well-being. If men relieve women of some household duties, then women will have more time for self-actualization and fulfillment. That may come in the form of employment, leisure time, sleep, or civic engagement, all of which bolster personal health and efficacy. In addition, allaying negative ideologies about agriculture in general, and women farmers in particular, will cultivate transformative and regenerative socioecological justice, or radical sustainability.

Agroecology is a regenerative agriculture that passively facilitates food-crop growth by engaging in a form of ecosystem democracy. Listening to and tending to ecological dynamics is fundamental to this form of agriculture. It connects human's labor to ecologies by requiring that people look, listen, and manage insect, plants, and bacteria populations. It is a collaborative relationship between human and nonhuman populations, whereby humans manage disease while ecosystem dynamics perform regenerative and reproductive labor that feed and sustain human and nonhuman populations. Gender divisions of labor in Cuba most often situate women in domestic spaces to regenerate and reproduce human farm laborers. However, my data show the regenerative contributions of women both in the home and on the farm. On the farm, bosses find them to be active contributors who outperform men at times. Divisions of labor situate women in observant roles where they are likely to see diseases or pests, and they are thus able to suppress crises before they occur. In addition, women have traditionally been the ones to prepare for potential natural disasters by preserving foods for times of crisis. With greater gender equity, men also may contribute to the life-saving work of disease management and disaster preparation.

Future research should build on this work by providing detailed analyses of how other social inequities interact with environmental outcomes in Cuba. In addition, it should specifically look at the environmental contributions and vulnerabilities of other traditionally marginalized groups, such as Black and LGBTQ Cubans, among others. Many questions remain about how political changes may affect Cuban food sovereignty. Specifically, after the death of Fidel Castro, if Cuba is currently becoming a protocapitalist country, then how might their commitments to social/ecological sustainability change? Will free-market imperatives, that bring out the selfish, competitive, and individualist aspects of people, prevail? Or, will we see Cubans create a subdued version of capitalism, where cooperation and competition remain balanced, and where human and ecological needs are sustained? Might Cuba create something socially and ecologically similar to the social democracies of Northern Europe?

Challenges aside, Cuba has made impressive strides toward socioecological justice and sustainability and they did so in a time of crisis. Terrible things could have transpired during the Special Period, such as famine, mass starvation, and civil unrest. But Cuba continued to support and nurture its human and limited environmental resources to come out of the crisis as a world-renowned leader in sustainable innovations. When people of the West succumb to similar socioecological crises, I hope that we all learn and adapt lessons from Cuba. Their socioecological resilience in the face of imminent threats is a model worth emulating. In the following chapters, I discuss the stories and values that contribute to Cuba's resilience.

4
Beyond Neoliberalism
The Promise of a Communitarian Story

> Thus, it can be seen as part of the long-term historical process by which the language of economics has come increasingly to colonise all area of social and cultural discourse.
>
> —David Studdert (2005, 35)

> Look at the legacy of poor Eve's exile from Eden: the land shows the bruises of an abusive relationship. It's not just land that is broken, but more importantly, our relationship to land. As Gary Nabhan has written, we can't meaningfully proceed with healing, with restoration, without "re-story-ation." In other words, our relationship with land cannot heal until we hear its stories. But who will tell them?
>
> —Robin Kimmerer (2013, 9)

Building stronger community relationships doesn't have to change everything we know. Socioecological community could include a synthesis of value systems that simultaneously appreciates individual expression and autonomy while also embracing community reciprocity and collaboration. (Indeed, hyper communitarianism can be stifling and oppressive to individuals who differ in any way.) However, as Greta Gaard states about environmental health and social justice, "as long as we do not address the conceptual bases for these forms of oppression and focus only on their symptoms . . . these pervasive structures of oppression will continue to operate" (2007, xxiv). Indeed, Jared Diamond (2005) contends that a society's survival requires long-term planning and reexamining core values when the old ones no longer serve. As evidenced by mounting ecological and social crises outlined previously in this book, capitalism's perverse incentive structure promotes fierce competition to grow short-term quarterly profits through

Surviving Collapse. Christina Ergas, Oxford University Press. © Oxford University Press 2021.
DOI: 10.1093/oso/9780197544099.003.0005

the dissection of ecosystems for commodities, or products exchanged for money and sold on the market. True wealth, turned into money, is human labor and the gifts of nature, and we have been ungracious beneficiaries, taking as much as we can and leaving pollutants in our tracks. This system is not serving most of us, and it never served the planet. But what are the values undermining societies? Which values should be discarded or replaced?

Neoliberalism at its core is a fragmenting paradigm that takes self-interest and individualization to their logical extremes. The atomization of humans, nonhuman species, and the biosphere serves as a divide-and-conquer strategy that breaks down whole systems into small vulnerable parts. These objectified pieces become more easily manipulated and controlled and thus exploitable. When organic agriculture was transformed from a movement about relational processes into an industry selling commodities (Guthman 2004), its most resilient features, such as polycultures, were rendered inert. The global economic system has severed many ecosystem dynamics, damming rivers that block salmon runs as an example, and is largely responsible for the myriad environmental problems we face. Strategies for social change must be multiple and varied and penetrate all dimensions, from socioecological structure to individual behavior. This chapter examines the limiting stories of neoliberal capitalism and considers how we might rewrite and restructure our stories to work toward a more just and sustainable society, or a radically sustainable economy.

4.1. The Ideas of the Ruling Class Are in Every Epoch the Ruling Ideas

During the Reagan Administration, a group of young lawyers were influenced by a book *Ideas Have Consequences,*[1] written by the conservative scholar Richard Weaver (1948). The title became their mantra as they helped create the conservative counterrevolution of the law (Hollis-Brusky 2015). Indeed, Pellow (2014) argues that ideas drive and sustain social movements and, in this case, countermovements. Discursive practices, such as dialogic writings and speeches, transfer ideas that influence cultural stories and values and are then reproduced and remade through individuals' interactions, codified rules and laws, and sanctions. Weaver's intellectual influence on the group of lawyers, who were then able to implement conservative laws, is a case in point. In this section, I examine neoliberal globalization, hyperindividualism, and the no-longer-viable supporting values and stories that perpetuate cultures of environmental degradation and social inequity. I do this so that we may begin to re-evaluate our current narratives and values, conceptualize other possibilities, tell new stories, and plan accordingly.

4.1.1. The Progress-Growth Paradigm

The hegemonic paradigm exported by the Global North is the neoliberal paradigm, based on old and new philosophies and economic rationales that influence policy decisions and corporate practices.[2] The theory of globalized sustainable development has been practiced largely under this framework. This paradigm is legitimized by institutionalized logics and is bolstered by cultural narratives and values that serve to self-perpetuate neoliberalism at the individual and interactional scales.

Culture describes fluid processes that weave together societal structures and individual agency. It is a historically embedded assemblage of power relations, meanings, values, traditions, lived practices, language, and stories that define spatial boundaries and affect how we understand and experience the social world. It also shapes our beliefs as well as behaviors. Values are shared cultural guidelines, or beliefs and ideals, about how individuals should or should not behave (Carruyo 2008). Cultural values are often expressed in cultural narratives and codified into rules and laws of organizations and the state, and they affect attitudes and how we respond to people and situations. While the concept of culture may seem nebulous, its supporting narratives do have material consequences.

Narrative and story are often used interchangeably, though they do have distinct meanings (Presser and Sandberg 2015). Stories generally refer to specific events, while narratives reflect broader cultural understandings (Presser 2008). Narratives, or stories, are a means of connecting individual experiences to larger cultural norms and themes. Stories are also "moral in nature" as "they pertain to right and wrong ways to act" (Presser 2013, 26). Eckstein (2003) argues that "storytelling . . . is about setting community boundaries, including some audience members within its territory and excluding others" (13). Narrators set these boundaries in the context of dynamic power relations, and understanding the social position and disposition of these narrators is important if we are to understand who these stories serve. However, Eckstein contends that "carefully told and carefully heard, stories do have the potential to act as a bridge between engrained habits and new futures, but their ability to act as transformative agents depends upon disciplined scrutiny of their forms and uses" (13). Indeed, Presser (2013, 25) contends that "stories make things happen because they guide human action." Stories are also fluid and change over the course of time.

Our political and economic institutions and embedded power relations reflect our hegemonic cultural narratives and values. In this way, the theory of political economic opportunity structure (outlined in chapter 2) has limited explanatory ability. Power relations humanize and assign moral rightness, compassion for, and legitimacy to power holders, while dehumanizing and devaluing those in subordinate positions (Presser 2013). Through discursive practices, these narratives legitimize power structures, inform our institutions, and are codified into rules and laws.

In particular, in the United States, the rugged individual is free from governmental scrutiny on their property. Americans understand property in the John Lockean sense as nature that a person has changed or manipulated in some way with their labor for its "betterment," or personal gain. This narrative informs laws about property rights, the right to bear arms, and warrants, searches, and seizures. In literature, such as nineteenth-century Horatio Alger pulp fiction, this same rugged individual is personally responsible for their lot in life and must pull themselves up by their bootstraps to rise from rags to riches. These stories also inform policy on welfare, healthcare, immigration, drug use, and innumerable others in the United States.

Our cultural narratives have material consequences on our economic system. We currently operate under a globalized economy that benefits a minority of elites who seek to commodify everything for their profit. This system is bolstered by stories that ever-growing displays of wealth, buying and owning more things, will make us happy, when in reality shopping feeds into loneliness. Stories of self-made millionaires encourage individuals everywhere to work harder, so that one day they may own all the world's riches. A secularized Protestant work ethic (Weber 1930; Conover 1975) and the recent rendering of the Prosperity Gospel (Van Biema and Chu 2006; Garber 2008; Rosin 2009) coalesce to instill values that convince people to work long hours in order to honor God, and the gifts for their devotion are earthly riches. In reality, these narratives benefit an elite who are able to exploit people's labor, and these earthly riches are natural resources, likely stolen from ecosystems and poorer people elsewhere. While some sects of Christianity, such as televangelists, feed people these messages, many of these stories are reinterpreted and recreated by a well-oiled marketing machine, paid for by the same elites that serve to make a profit off of the commodification of everything.

The current global paradigm is proving unsustainable and unjust. What is needed is a paradigm shift, or something of a new ecological paradigm, the likes of which sociologists have spoken of since the 1970s (Catton and Dunlap 1978). This new ecological paradigm centers the relationship between human and more-than-human nature and, rather than treating them as separate, treats them as mutually constitutive. Before we can understand this new paradigm and its scaffolding, we must first understand the values of the unsustainable dominant neoliberal paradigm.

4.2. The Neoliberal Paradigm

Capitalism is an economic system predicated on economic growth, often measured by gross domestic product (GDP), with vague assumptions about progress and development. Capitalism is bolstered by stories of the moral correctness of inequality, ownership, and wealth accumulation, as well as deserving and undeserving property owners. Based on classical economics, a new form of capitalism emerged in the 1970s and '80s that many scholars and analysts call neoliberalism, which takes individual competition and unencumbered free markets to their logical extreme. However, this is all done under local-to-global social hierarchies that stack the deck in favor of those already among the elites. In this section, I highlight some of the stories and values associated with neoliberalism.

4.2.1. Liberal Individualism

Before exploring neoliberalism, it is important to understand the notion of individualism. Individualism is an underlying ideology of liberalism that emphasizes the individual. Self-reliance, personal responsibility, and independence are essential values of individualists, who believe that individual interests, desires, and goals should take precedence over the goals of a society, social group, or government (Wood 1972).

The importance of individualism to neoliberalism was made explicit in 1987 when Margaret Thatcher, then prime minister of England, famously stated while talking about social safety nets:

> There is no such thing as society. There is living tapestry of men and women and people and the beauty of that tapestry and the quality of our lives will depend upon how much each of us is prepared to take responsibility for ourselves and each of us prepared to turn round and help by our own efforts those who are unfortunate. (Thatcher 1987, 30–31)

She spoke of the individual as the primary social unit in a nation. During her term as prime minister, Thatcher established national policy and, with the help of other international organizations such as the World Bank and International Monetary Fund, and world leaders like President Ronald Reagan, an international movement that systematically cut social safety nets and promoted an ideology of personal responsibility. It's worth noting that individuals are politically inferior entities because, unless they are wealthy, individuals have limited resources to be of much influence.

4.2.2. Neoliberalism

Neoliberalism is a political movement and approach to economics and governance that began influencing nation states in the 1970s. It extends individualism into the market and its goals are growth and accumulation of wealth. Based on similar ideologies as old liberalism and classical laissez-faire economics that saw no role for the state government and instead presumed market-driven governance the most efficient, debates about neoliberalism arose in economic circles in the early twentieth century. These debates gained new strength after the old liberalism wrecked the US economy and sent it into the Great Depression. This new liberalism was meant to add a minimal role for the state to protect private property, offer some poverty alleviation, and maintain social order. Thus, neoliberal state governance is reconceived as a public-private partnership. However, many argue that the neoliberal state is in fact a powerful one that wields militarized social control when anyone subverts corporate or private interests. The modern neoliberal society allows the market to govern with state protections, and the market logics of competitiveness, efficiency, and profitability determine how social goods are distributed. It turns individuals into "entrepreneurs in their own lives" who are wholly responsible for all of their failures. Some neoliberal policies include a dogmatic veneration of economic growth, austerity (eliminating social welfare programs), privatization of public services and goods, and economic liberalism (individual competition, free trade, and private property), along with lax business regulations that amount to a flagrant disregard for negative social and environmental externalities—that is, the consequences of commercial or industrial activities that are not reflected in the cost of goods and services, such as pollution or negative health outcomes (Bockman 2013; Thompson and Schor 2014).

Friedrich von Hayek, a famous economist who advised Margaret Thatcher and is known for his laissez-faire economic views, tended to be more suspicious of community. His views aligned with historical philosophies of communities and communitarian ideologies as socially stagnant arrangements that stifle personal liberties through top-down, authoritarian rule (Bockman 2013; Thompson and Schor 2014). In turn, he believed that rather than exchange relationships—based on social capital, trade, or barter—markets were the most efficient socioeconomic mediators between self-interested, individual free agents, and that price was the most adequate reflection of supply-demand relationships. Because of private property ownership, individuals possessed rational information and incentives that allowed them to make efficient decisions about their property and to obtain more of it. In this view, social harmony between diverse self-interested individuals is maintained by local market mechanisms rather than community bonds and social solidarity. Indeed, competition between these individuals is what delivers the best quality goods at the lowest price, according to this logic. This view ignores what Mies so aptly articulates, "the core of individual freedom . . . linked to money and property, is the self-interest of the individual and not altruism or solidarity; these interests will always compete with the self-interests of others. Within an exploitative structure interests will necessarily be antagonistic" (Mies and Shive 2014, 66–67). This antagonism pits would-be allies against each other, as one person, a worker, desires higher wages and another person, a consumer, desires lower costs.

The "survival of the fittest" narratives bolstering neoliberalism and neoclassical economics are based on an outdated misunderstanding of Darwin's theory of evolution. They emphasize a worldview of fierce competition for scarce resources between selfish individuals. The fittest entrepreneurs survive and are handsomely rewarded for their efforts, while all other businesses perish. However, evolutionary theory has advanced quite a bit since these economic misinterpretations of Darwin, and scholars have since recognized and advanced the importance of cooperation and symbiosis in evolutionary processes (Nowak 2006; Hrdy 2009; West, El Mouden, and Gardner 2011). Thus, the structure of capitalism in the neoliberal era is situated on a misunderstanding of the philosophical and science-based analogies of evolution, and furthers a Hobbesian, *Leviathan*-esque, competitive lifestyle that is "nasty, brutish, and short." These narratives promote underlying values that further an economy based on the "competitive struggle between atomized individuals" (Thompson and Schor 2014, 234). In this case, the ends justify the means, and the means for survival are justified no matter the cost.

4.2.3. Violence as Necessary and Just

Neoliberal economists often ignore the fact that "survival," or in this case political and economic hegemony, is maintained through a powerful militarized state, or imperialism. The history of the United States is a violent one. In search for more resources, Western Europeans set the stage through the colonization of Indigenous peoples and lands before the inception of the United States as a nation. The US government continued this legacy through the enslavement, genocide, displacement, and forced assimilation of native peoples. Contemporaneously, industrialization was largely built on the backs of enslaved peoples and indentured servants. Slavery became contentious in the United States around the time when the need for human labor was curbed by mechanization and fossil fuels. It was only when the energy from fossil fuels, then an abundant and cheap source of energy, could be easily and economically harnessed that slavery ended in the United States. The rest of necessary labor could be coerced inexpensively from the previously rural poor. This violent paradigm has led to hundreds of years of wars, most of which have *not* occurred on US soil.

In fact, the global food system is supported through research and development done specifically for war. The Green Revolution constituted the beginnings of industrial agriculture during the mid-twentieth century, largely after World War II. This period marked a shift in agriculture toward shrinking rural farm populations, concentrated large-scale farming, mechanization, and the use of fossil fuel–derived synthetic fertilizers and pesticides (Pfeiffer 2006; Shiva 2016). After WWII ended, chemical companies had an abundance of unused chemicals developed specifically for chemical warfare, such as nerve gas. Some of these chemicals had the added benefits of being effective insecticides and accelerating the nitrogen fixation process required in agriculture. Specifically, the Haber-Bosch process for synthesizing ammonia from nitrogen and hydrogen gases was instrumental for creating both explosives and fertilizers (Pfeiffer 2006; Shiva 2016). This process was simultaneously responsible for the expansion of agricultural production and concurrent population growth as well as chemical warfare and the loss of many lives.

Global North hegemony and affluence is maintained by what is in effect a perpetual war machine, termed the military-industrial complex because it is an alliance between military, industry, and government policy. Perpetual war not only serves to maintain United States' access to resources from all over the world but also causes environmental destruction in and of itself (Jorgenson and Clark 2009; Jorgenson, Clark, and Kentor 2010; Downey, Bonds, and Clark 2010; Alvarez 2016). The United States has been at war for over 90 percent of its history, has about eight hundred military bases in more than seventy countries, outspends any other nation on its military budget, and has launched the most military operations since WWII (Wiist et al. 2014; SIPRI 2017).

This, in combination with the fact that the United States consumes more per capita in many resources, could lead some to question the nation's motives. According to Sierra Club's Dave Tilford, as quoted in *Scientific American* (Scheer and Moss 2017):

> "With less than 5 percent of world population, the United States uses one-third of the world's paper, a quarter of the world's oil, 23 percent of the coal, 27 percent of the aluminum, and 19 percent of the copper," he reports. "Our per capita use of energy, metals, minerals, forest products, fish, grains, meat, and even freshwater dwarfs that of people living in the developing world."
>
> He adds that the US ranks highest in most consumer categories by a considerable margin, even among industrial nations. To wit, American fossil-fuel consumption is double that of the average resident of Great Britain and two and a half times that of the average Japanese. Meanwhile, Americans account for only four percent of the world's population but create half of the globe's solid waste.

It would almost seem as though the United States is trying to protect more than its world domination or global hegemony, but also its unequivocal access to all of the world's resources. The stories that glorify American exceptionalism and encourage belief in the Global North consumer lifestyle also perpetuate neocolonialism and legitimize violent defense of that lifestyle no matter the cost. It's a story that reduces all natural systems to their instrumental use value for Northern elite humans. Northern lifestyle treats all of nature as objects that fulfill human desires and needs, not as having value in and of themselves. This story promotes a belief in humans' superiority over nature, which grants us the right to own the raw materials of nature and to defend that ownership.

The beliefs in moral superiority and right to ownership are the first acts of violence that justify all the rest. As an expression of this violence, European settler colonialism began in the Americas in the fifteenth to sixteenth centuries and was upheld into the coming centuries by narratives of Manifest Destiny—or the nineteenth-century belief of settlers' rightful westward expansion across North America—and *terra nullius*, or legally "unoccupied land," which legitimized the displacement and genocide of Indigenous communities throughout the continent (Holleman 2018). This violence may be rooted in a longer or more deeply entrenched history than colonialism or the resultant expansionist US policies. This story, as Diamond suggests, might be as old as agriculture, or perhaps even warring tribal societies (Keeley 1996). Although the origins of this violent paradigm might be difficult to ascertain, other extremely coercive forces, such as public relations and marketing, keep affluent individual consumers in the Global North complacent and unaware of the violence that people and ecologies, termed "sacrifice zones" (Lerner 2010), endure elsewhere.

One expression of markets governing with state protections is the US response to climate change. A nefarious form of elite environmentalism is emerging, not unlike the Malthusian sentiments or social Darwinism of the 1800s that saw the poor as immoral, inferior, weak, and undeserving of power or help, especially in relation to scarce resources. Many fear, as Klein (2014, 54) suggests, that as the environmental crisis increases in severity it may become "the pretext for authoritarian forces to seize control in the name of restoring some kind of climate order." Some have termed this eco-fascism or "green fascism." Personal responsibility rhetoric has callously turned to "victims deserve their fate" (Klein 2014, 53), as white supremacy makes a comeback among conservatives. In the name of "national security," the state is protecting the white elite through draconian immigration policies, such as Trump's border wall and migrant detention centers, as a response to a surge of asylum seekers fleeing internal conflicts in their own countries. Migrants from Guatemala, Honduras, or El Salvador seek refuge as a result of violence and poverty as well as persistent droughts due to climate change. The irony is that the poorest communities in both developed countries, such as New Orleans during hurricane Katrina, and less-developed countries, experience the worst effects of climate change even though they've contributed the least greenhouse gases (GHGs).

Poor communities also have the least resources to adapt; thus, the UN (2019) is projecting "climate apartheid" into the coming decades as the private sector spearheads community resilience and disaster preparedness public works. The affluent have more available money and resources to prepare their communities for disasters; they also tend to live on the most desirable land that is least disaster prone, such as on higher ground away from flood plains. Affluent individuals also have resources to rebuild after disasters. The opposite is true in every respect for poor communities—they have few resources to prepare for disasters, live in the most disaster-prone areas, and have fewer resources to rebuild after disasters strike. Without transformative change toward social justice, the UN Human Rights (2019) report recently warned:

> Rather than helping the world adapt to climate change, privatizing basic services and social protection may be a form of maladaptation. When hurricane Sandy wreaked havoc in New York in 2012, stranding low-income and vulnerable New Yorkers without access to power and healthcare, the Goldman Sachs headquarters was protected by tens of thousands of its own sandbags and power from its generator. Private white-glove firefighters have been dispatched to save the mansions of high-end insurance customers from wildfires. "An over-reliance on the private sector could lead to a climate apartheid scenario in which the wealthy pay to escape overheating, hunger, and conflict, while the rest of the world is left to suffer." (2019, 14)

Neoliberalism's reliance on the private sector to care for pressing social problems continues to leave many behind, and the US government uses violent means to keep the most vulnerable at bay. However, when physical violence proves counterproductive among productive and consuming Westerners, other more insidious forms of social control are leveraged.

4.2.4. How the Power Elite Gaslight the Rest into Quiescence

Gaslighting is a term used in psychology to describe a type of abusive behavior. The term was popularized by a play, later a movie, entitled *Gaslight*, wherein a man attempts to convince his wife that she is going insane by flickering the gaslight in their home. When his wife notices the flickering, he persuades her that she is in fact imagining it. He further lies to her about a number of things, such as objects mysteriously disappearing and reappearing, manipulating her into questioning her own perception. Ultimately, he is doing this to have his wife committed so that he can seize her property and hide the fact that he murdered her aunt. Gaslighting then is the "effort of one person to undermine another person's confidence and stability by causing the victim to doubt his or her own senses and beliefs" (Kline 2006, 1148). An abuser usually gaslights their partner in order to deflect attention from the abuser's transgressions while simultaneously attempting to control the victim's thoughts by making them doubt their own memory and perception.

Historically, psychologists used the term gaslighting to refer to intimate relationships; however, some have begun to apply the term to sociopolitical and historical processes and institutional practices. Burrow (2005) engages the power that dominant groups' dismissal of oppressed peoples' emotions has by focusing on feminist anger, which she calls a subversive response to oppressive practices. She notes "dismissal silences one's political voice and, at the same time, compromises a valuable source of self-worth and self-trust" (27). Davis and Ernst (2017) extend this and apply the concept to white supremacy, arguing that "racial gaslighting—the political, social, economic and cultural process that perpetuates and normalizes a white supremacist reality through pathologizing those who resist," helps to maintain white supremacy in the United States. I use the term more broadly to include the myriad ways structural dominance is ideologically maintained, including but not limited to heteropatriarchy, white supremacy, speciesism, and class hierarchy. Davis and Ernst argue "this manipulation of perception is powerful because our reality—how we perceive the world and our place in it—is socially constructed." Gaslighting in this way "relies on the production of particular narratives" (Davis and Ernst 2017, 2–3). Marx wrote of these narratives in *The German Ideology*. He put forward the "ideas of the ruling class are in every epoch the ruling ideas." Elites have an interest in maintaining master narratives that legitimize their superiority and perpetuate myths that oppressed peoples are at fault for their oppression because of their own moral failings. Examples of this include mythologizing the lazy poor who don't work hard enough to deserve wealth; blaming survivors of sexual assault because they "dress too provocatively"; or blaming Black people for their disproportionate arrests or even fatalities because they should respect cops and remain compliant. Sociopolitical gaslighting includes tactics such as "diverting issues from legitimate targets" and instead changing the focus (e.g., tone policing; Burrow 2005, 31); victim blaming; restricting freedom of expression; feigning innocence; and outright dismissal of victims' claims.

In the United States, the political economy functions much like an abusive relationship. Political, social, cultural, and economic institutions all work in service to the economic goals of growing GDP. Scholars of gender and race recognize all of these processes as at once gendered and racialized, defining capitalism as heteropatriarchal and white supremacist in origin, and perpetuated through current institutionalized practices (Acker 2006; Pulido 2017). In an effort to fulfill the economic growth imperative, our economic institutions attempt to maintain cheap and compliant labor. Historically, this meant divisions of labor whereby women performed free domestic and reproductive labor for their wage-working men, people of color were enslaved, and working-class white men received low wages and worked long hours. These free or cheap forms of labor subsidized, and continue to subsidize, the capitalist economy. As slavery became illegal throughout the world, although it persists, other less deplorable forms of coercion became necessary (Walk Free Foundation 2018).

Resistance to oppressive structures is pathologized. As an anecdote, the DSM-5, a psychological diagnostic manual, includes an actual psychological diagnosis in children called oppositional defiant disorder that includes symptoms such as defying authority, among others (American Psychiatric Association 2013). But there are more serious examples still. Ample platitudes remind us to be grateful to have a job at all and to not "bite the hand that feeds." In addition, people who resist are often dismissed as "lazy" people who don't want to work or are called names like "utopian idealists." Environmental activists are often dismissed as "tree-huggers" or "hippies." If you become depressed or anxious as a result of your political disempowerment, cultural stories would have you believe that you are mentally weak. The cult of positive thinking insists that you can "choose happiness" by remaining grateful for *what you do have* and not focusing on how your life might be better. Everything will be fine if you just "stay positive" enough. Indeed, many will remind you that "only you can make you happy" (Ehrenreich 2009). Moreover, pop psychology suggests that you suffer from a "chemical imbalance" and need medication to keep you numb and compliant. And so, when someone realizes that things aren't right, when they begin to question the way things are, they are told that they are being difficult—that they are the problem—not the structures that exist! They are gaslighted into quiescence. As a result, many are medicated to cope with feelings of helplessness. The media of the affluent West also do their part to keep people complacent, consuming, and ill-informed.

4.2.5. Consumer Society, Marketing, and Social Isolation

A marketing-perpetuated myth that deeply permeates many societies is that more consumption will make us happy. In reality, as consumption has escalated in the last sixty years, mental health has steadily declined. In the United States, loneliness has doubled since the 1970s and '80s, with 35 percent of survey respondents reporting that they are lonely (McPherson, Smith-Lovin, and Brashears 2006; Wilson and Moulton 2010). Loneliness is associated with premature death, poor physical and mental health, and drug addiction, as well as increased social media usage in lieu of face-to-face interaction (Cacioppo and Patrick 2008). More people live alone than in previous decades and have fewer confidants (McPherson, Smith-Lovin, and Brashears 2006). Chronic loneliness is also associated with materialism and shopping, and higher materialistic values are associated with lower prosocial values (Pieters 2013). In short, the more we value conspicuous consumption, the less altruistic and less connected we are to others, and thus the less likely we are to be concerned about the well-being of others.

While rational choice theory in economics assumes that individual consumption choices are predictably based on how they might maximize their individual happiness or utility, marketers have learned how to manipulate this process. Exploiting individuals' deepest insecurities, fears, and desires is at the heart of modern marketing. This technique was honed by Edward Bernays, deemed the father of public relations and named one of the most influential Americans of the twentieth century by *Life* magazine (1990). Throughout the century, Bernays based his practice on crowd psychology and psychoanalysis. He believed that "the masses" were irrational and needed to be controlled for their own good. Fortunately, according to him, the masses are easily manipulated by tapping into their innermost fears and desires (Bernays 1928). He argued that a small minority of elites should and do control and shape ideologies, political beliefs, and even which items people should buy. His tools and techniques are still widely used today.

Based on the Bernays tradition, researchers argue that marketers feed off insecurities (Kilbourne 1999) while simultaneously sending "meta-message[s] that you can solve all of life's problems by purchasing the right products" in order to keep people consuming (American Psychological Association [APA] 2004, 60). Indeed, the APA has noted that consumerism is tied to loneliness, insecurities, and lower life satisfaction (DeAngelis 2004). Marketers are aware that there is a twinge of happiness that comes with purchasing new things, termed "retail therapy," and this short-lived phenomenon drives the insatiable hunger of consumerism (Atalay and Meloy 2011; Rick, Pereira, Burson 2014). Marketers exploit our needs to experience novelty and intimacy in their ad campaigns. However, coping with loneliness by acquiring new things will never satisfy us, because as Kilbourne (1999) points out, things can't love us back.

Today we more often live alone, are more socially isolated, and feel lonelier than in previous generations. How did we get here? What were the social mechanisms that drove us to such extreme isolation, and, in turn, such extreme consumption? The factors are probably numerous, but much is likely linked to the nature of hyperindividualism characteristic of neoliberal globalization.

4.3. The Origins of Human Exemptionalism and the Control of Nature

In Litfin's book on ecovillages, she describes the most compelling story of Western society as the "conquest of nature through technology" (2014, 4). Westerners are able to tell this story because of preexisting stories that presume humans' ability to separate themselves from the whole of nature, or human exemptionalism (Catton and Dunlap 1978). Religious notions of humans' elevated status in relation to nature as well as their moral claim to dominion over nature further humans' separation and justify conquest. Because Western culture assumes that humans and nature are separate and that humans are superior, many societies use technologies to protect themselves from the capriciousness of nature, buffering the uncertainty with creature comforts and conveniences. Humans have not always seen themselves as outside of or separate from nature. This is a relatively new story in human history, with origins in some Greek philosophies, Christianity, and the Scientific Revolution (Merchant 1980).

4.3.1. Controlling Nature

In *The Death of Nature,* Carolyn Merchant (1980) discusses the controlling imagery that different societies use to identify or characterize nature. Controlling images of people or things place ethical restraints on practices and define how we should relate to them. She argues that historically there were two competing images of nature that arose from organic cultures. One story of nature was that of nurturing mother, a benevolent earth that provided sustenance for all creatures' needs. The alternative story saw nature as a violent and chaotic feminine force that sent storms, droughts, and disasters to innocent humans. Merchant contends that these worldviews carried value systems that had direct effects on gender hierarchies and practices toward the land.[3] Nature as disorder began to predominate in Europe as the Scientific Revolution of the sixteenth century took hold, and a new ethos evolved that sought to dominate and control nature through mechanization. She writes, "As Western culture became increasingly mechanized in the 1600s, the female earth and virgin earth spirit were subdued by the machine" (2). The ideal that humans held moral superiority and therefore dominion over earth existed in Greek philosophy as well as Christianity, but the Scientific Revolution took this story and spread it through social, economic, and political spheres.

The Scientific Revolution put forward a new mechanistic worldview that sought to understand whole bodies and systems by picking them apart, favoring individuation. Shiva (2016, 4) posits the Scientific Revolution as Newtonian and Cartesian; she quotes Newton himself as writing "solid, massy, impenetrable, moveable particles . . . are so very hard, as never to wear, or break in pieces: no ordinary power being able to divide, what God himself made one in the first creation." Each atom, particle, or component was dissected and held under a microscope for scrutiny as disassociated parts of a machine, through analogy. The machine analogy conjures images of coldness, lifelessness, passiveness, and inactivity. This image is projected onto nature as a passive vessel that requires an active man to manipulate and control it toward his ends. Thus, to man, the most important aspects of nature are the use value or utility of "her" parts. The Scientific Revolution saw the old narrative of erratic woman in need of containment converge with a new narrative of nature as machine-like and therefore passive and manipulable. Either image justified the by-any-means necessary and indifferent approaches that the West assumed. From there, acts once seen as callously "raping" mother earth's virgin soils and forests, like mining and nonregenerative logging, became legitimate and necessary forms of control to force natural systems into submission and to meet the needs of humans. Indeed, these acts were considered "improvements" made to the land, which further justified control and ownership, or private property (Merchant 1980).

Narratives can be contradictory, and the more familiar the story, the more acceptable its inconsistencies (Presser 2013). Organic societies' early fears of nature were well founded. Natural disasters such as flooding or droughts could bring famine and disease. Volcanic eruptions or earthquakes could devastate or destroy whole communities. Thus, wanting to feel a sense of control and security in the face of these potential disasters is understandable (Merchant 1980). However, as if in a sleight of hand maneuver, human societies projected their own vulnerability and insecurities onto nature, and turned nature into the vulnerable and passive receptor of humans' rapaciousness.

Work on moral exclusion, or the act of perceiving some groups of people or entities as "outside the boundary in which moral values, rules, and . . . fairness apply," indicates that many people preclude the possibility that nonhuman animals and nature can even experience harm (Opotow 1990, 1; Opotow 1994). Rather than recognizing intrinsic worth and protecting nature based on that alone, humans are more compelled to conserve "natural resources" because of their utility (Opotow 1994). Others similarly argue that commodification, or the sale or exchange of products on the market, further concretizes this relationship to nature (Longo, Clausen, and Clark 2015). Humans see themselves as separate from nature, and, by God-given right, able to own land or nonhuman animals, and, moreover, they are incentivized to extract what they please for profit. This story further alienates, atomizes, and severs connection between humans and the more-than-human world. Moral exclusion is also integral to other social inequities. It is how societies come to see some people and their labor as raw materials, naturally tied to the land and legitimately paid less, such as migrant farm labor in the United

States. Institutional, physical, cultural, social, and psychological distance maintain barriers between those who are deemed worthy of property and wealth and those who are not (Opotow 1990). And we are back to the individualist origins of the neoliberal story.

4.3.2. The Human-Nature Dichotomy Deconstructed

There is evidence to suggest that many societies have taken the nature/civilization dichotomy too far, and this story is hurting both people and the environment. Emerging research shows a relationship between green space and social cohesion. This research indicates not only that individuals experience improved psychological well-being when they are exposed to nature or green space (Doherty 2016), but also that this in turn makes them feel more connected to place and the surrounding community, and therefore less likely to commit crimes. Connection to nature also positively affects environmental stewardship. People who do not have access to green space feel less happy, are less physically healthy, have fewer community connections, live in less equitable environments, and express less concern for the environment (Weinstein 2015; Adams and Jordan 2016; Doherty 2016; Uzzell, Pol, and Badenas 2016). A positive feedback loop can be established in either direction. Because of neoliberal globalization, we appear to be stuck in the more destructive feedback loop.

4.3.3. A Death Machine That Feeds on Death

We are living through what scientists have termed the sixth mass species extinction, caused by humans (Ceballos, Ehrlich, and Dirzo 2017). Destruction of habitats brought about by extraction and mining, deforestation due to agriculture, pollution from industry, ocean acidification, and climate change are some contributing factors. Economic development is responsible for these large-scale land-use changes. Vandana Shiva (2014), a critical scholar of development, has termed our economy a "necroeconomy—its profits are rooted in death and destruction" (7). She argues that this economy is based on laws of exploitation and domination that reduce most life, human and nonhuman, into raw materials and inputs for commodity production and ultimately profit. It sees value only in people and nature that can be exploited, and its primary energy source is the inefficient fossil fuel.

The fundamental source of energy for almost all life on our planet is the sun. Plants, algae, and some bacteria are able to process sunlight for fuel. Plants also consume soil nutrients. Herbivorous animals obtain their energy by consuming plants that store solar energy and soil nutrients, and carnivorous animals consume animals who have stores of plant energy. Animal waste from this process returns nutrients to the soil that plants then consume; this is a closed-loop cycle (Pfeiffer 2006). Shiva (2016) refers to this regenerative life cycle as part of the law of return "where there is no waste; everything is recycled" (xi), and it provides nourishment.

Fossil fuels, on the other hand, are derived from decayed organic matter buried in the earth's crust and exposed to heat and pressure over hundreds of millions of years. Once living plants and animals consuming solar energy, fossil fuels are the remnant, decomposed solar energy of the dead (Pfeiffer 2006). Fossil fuels are a far less efficient form of energy than the sun. Solar-powered agriculture consumes four kilogram-calories (kcal) of energy for one kcal of food energy, whereas fossil fuels require ten units of energy to produce one unit of food. Fossil fuels also generate waste in the forms of GHG emissions, toxic soil, and water pollutants (Pfeiffer 2006; Shiva 2016). Capitalism has been feeding off this dead energy since the early days of the Industrial Revolution, which began in the 1700s.

In essence, capitalism takes life and, with dead fuel, turns it into death for the market. Timber companies only see value in the forest as lumber and cut down trees to sell wood. The ecological benefits of a forest—such as maintaining biodiversity, preventing soil erosion, controlling climate, sucking carbon out of the air—have no economic value. Thus, there is no incentive to maintain the integrity of the forest's ecological system. The economic rational of capitalism opens closed-ecological loops by prioritizing its own regenerative profit and growth cycle, its profit-loop, through a linear consumption and exploitation of ecosystems. The commodity production process takes whole systems, such as forests, and fragments and objectifies them, turning them into products for profits, such as timber (Mies and Shiva 2014; Shiva 2016). Whatever materials are not useful in the process are discarded, wasting nutrients and turning them into pollution. Feedlots pose another example of this dislocation. While it is economically rational to condense all cattle into feedlots and feed them overabundant, cheap corn, which fattens them up more quickly than grass feeding, it is ecologically inefficient to do so. Because the feed in feedlots consists of processed corn and pesticides, cattle are more likely to become ill, necessitating antibiotics and contributing to the growth of antibiotic-resistant bacteria. In addition, because of their feed, cattle's manure becomes toxic and unusable as fertilizer, and therefore turns into pollution, expelling methane into the atmosphere or potentially polluting waterways (Pollan 2006). Conversely, on a farm that uses rotational grazing, animals and crops coexist, and the cows' waste is used to fertilize crops (Shiva 2016).

Other examples of sacrifice zones for capital abound. In industrial agriculture, farmers overspray crops with synthetic-chemical fertilizers in order to sell more and earn more money. Crops fail to consume all of these fertilizers. The nitrogen and phosphorus run off into streams, rivers, and eventually the ocean where they cause eutrophication, or overnourishment of water-based plant life. This can lead to algal blooms that suck oxygen out of the surrounding water, creating literal dead zones where other aquatic life suffocates to death. The application of pesticides in industrial agriculture also causes harm to many species that are not pests. Pesticides indiscriminately kill even beneficial microorganisms, such as fungi, earthworms, and bacteria, in the soil that provide nourishment to plants and maintain soil integrity, preventing soil erosion and fixing nitrogen. Certain pesticides kill important pollinators, like Monarch butterflies and bees, that do the necessary work of pollinating crops. Neonicotinoids in particular are cited as one of the main causes of bee colony collapse disorder. There are many instances of such rifts in capitalism where economic efficiency is promoted over ecological efficiency, which are outlined in previous chapters (Foster 1999; Pollan 2006; Shiva 2016).

The scaling up of the organic food production process provides another illustrative example. Before USDA certification, organics began as a varied set of practices that had guiding principles based on closed-loop, systems-approaches to regenerate soil fertility. Specifically, regenerative processes include nurturing a diverse ecosystem by harnessing solar energy and preserving soil through low-tillage methods, cover crops, polycultures, green and animal manure fertilizers, and biological pest reduction systems that rely on maintaining beneficial-to-pest insect ratios. This type of agriculture requires more ecological and human labor than heavy machinery. In the process of narrowing the definition of organics to create replicable industry standards, many of the original activist visions of organics were lost, including environmentally sound practices and social justice considerations.

The certification process dismissed the systems-approaches to organics, like agroecology, and organics were reduced to a predefined set of acceptable inputs, like nonsynthetic pesticides and fertilizers. What once was a set of relational practices based on maintaining a biodiverse and dynamic ecosystem became a monoculture of suitable nonsynthetic inputs. USDA certification allows the use of monocropping, employing underpaid migrant labor, mechanizing post-harvest processing, and purchasing organic inputs (Buck et al 1997). As a result, the rise in certified organic production is associated with increases in GHG emissions and increases in water pollution from the use of fertilizers (McGee 2015; McGee and Alvarez 2016). In large-scale agriculture, biodiversity dwindles as weeds and insects are killed by pesticides and herbicides, and heavily fertilized monocrops thrive. Monocultures are made possible by abundant synthetic fertilizer use and fossil-fuel guzzling machines, including tractors, seeders, harvesters, and irrigation pumps, that require little thought or labor (Rogers 2010a, 2010b). Thus, organic regenerative ecological processes were not scaled up for commodification. Instead they were reconfigured as a mechanistic, linear, lifeless process that kills anything unprofitable and runs on dead energy, or fossil fuels.

There is little incentive to conserve ecosystems like forests because capitalism is a system that thrives on scarcity, as abundant resources are cheap, and scarce resources garner more value (Kimmerer 2013). Thus, capitalism rapaciously kills to create scarcity to garner higher profits while feeding on dead energy. But, there are lively oppositional struggles against the carnage; as the activist saying goes, "respect existence or expect resistance."

4.4. The Co-Optation of Plentitude and Rhizomatic Change

In *Sustainable Lifestyles and the Quest for Plentitude*, editors Thompson and Schor (2014) make the case for "rhizomatic resistance," which activists have advocated for decades (see Kellogg and Pettigrew 2008). The rhizome is often used as a metaphor for grassroots resistance and refers to plants that grow vast networks of roots underground that allow them to laterally self-propagate, such as certain grasses. Thompson and Schor (2014) apply the term to their vision of an alternative economy, the "plentitude economy," that is emerging from the grassroots. "Rather than looking to top-down government solutions, the plentitude paradigm suggests that at the current moment, the diversified actions of entrepreneurial agents generating new assemblages of technology, human capital, lifestyle practices, and market-mediated social relationships are the more likely impetuses for a more sustainable and emotionally rewarding economy" (234).

Schor and colleagues argue that what is needed is more of an emerging eco-habitus. Borrowed from Bourdieu, habitus is meant to describe socioeconomic class-based tastes or dispositions (Carfagna et al. 2014). Eco-habitus describes collective consumption strategies that promote ethical consumption, a type of green consumerism that acknowledges fair labor practices. Their research finds a growing number of high cultural-capital consumers, generally the middle to upper class, who are basing consumption decisions on environmental considerations (Carfagna et al. 2014). These grassroots consumer-activists are presumably changing the economy with their prudent purchases. It makes sense for researchers to recommend some kind of environmental consciousness and changing of social values, and it is important for affluent consumers to make better decisions and to consume less. However, the emphasis here is still on markets and consumption, rather than ecologies and social relationships. This research also serves to perpetuate the myth that only the rich can afford to be environmentally responsible. Socioecological problems stem from the cultural primacy of tending to economics before tending to social and environmental needs and connections. The Earth and its inhabitants would be better served if people focused less on consumption and more on each other. Building stronger relationships, and in turn, communities by interacting with the nonhuman natural community is one way to think about this.

Rhizomes are important parts of ecosystems and are compelling imagery for a sprouting, underground, radical movement. Yet, while rhizomes are an essential and literally growing part, they are just that, a part. Rather than picking apart the pieces of our societies and systems, it is important that we remember that each piece is a part of a whole that needs attention at every scale. Without attending to our global economic system and working to change it, our small-scale grassroots initiatives will do little to alleviate the problems at hand. While I see the utility in a large-scale movement of disengagement and divestment, we must work directly to disrupt current economic practices, while also focusing on the needs of our human and biotic communities. Recognizing the class privilege inherent in eco-habitus, Holt (2014) suggests making green jobs the only option, which would require large-scale change. Neoliberalism is a structure adept at co-optation. Indeed, some of the most subversive acts, like sharing cars or tools rather than buying them new, are commodified in this system.

4.4.1. Neoliberal Co-Optation of Oppositional Stories

Scholars and activists are calling for new economic systems and lifestyles in the Global North that include self-provisioning, downsizing, and sufficient fulfillment of material needs in opposition to the excesses of consumer lifestyles. While individuals may try to consume less, consumption is still at the center of this framework. It is true that the affluent of the Global North need to succumb to the realities of "plentitude" (Thompson and Schor 2014) or "sufficiency" (Wichterich 2015). However, this ignores the material reality of the impoverished who struggle to make ends meet and lack the luxury of consumption in the first place. Indeed, plentitude consumption still requires some social capital, often is more expensive because of its smaller scale, and doesn't speak to most of working-class middle America (Holt 2014). Further, this perpetuates the idea that in a consumer society, the only legitimate environmentalist is an affluent person who can afford to own an electric car and buy organic, or who has the leisure time to turn their compost. In the United States, much of what is called environmentalism is actually about consumption. This positions a poorer person who may own an old gas-guzzling car as anti-environmental, although this same person may be unable to afford new clothes for their children or even enough food to feed them. What is needed is a holistic approach that will make environmentalism materially viable for the poor and affluent alike. Rather than manipulatively marketing environmentalism (Holt 2014), holistic environmental livelihoods must become the new normal, and not narrowly conceived of as green consumption, because in a neoliberal context consumption practices of any kind are readily transformed into niche markets.

Editors Thompson and Schor (2014, 234) showcase an array of new economic relations sprouting up all over the world, many based on deep-green consumption. Of these economic relations they concede, "Family resemblances do exist between the neoliberal valorization of competitive markets and the market practices that characterize plentitude-oriented economies, but they are embedded in very different ideological systems." From this I emphasize, "different ideological systems" embedded in a neoliberal system. I caution against such excitement about these new economic relationships because, as Bockman notes, "the collective goods and experiments in living that we create together are appropriated by corporations to create new conditions for profit and capitalist power" (2013, 14). Indeed, Holt suggests that these new economic movements "pose no systemic threat to BAU," or business-as-usual economics, and rather, have "been readily incorporated as a valued new cultural source material for many thousands of new businesses" (2014, 210). I argue similarly that oppositional ideological and value systems are necessary. But, rather than prioritizing easily co-opted new economies, the focus should turn to nourishing socioecological relationships. As we'll see, within the context of neoliberalism, these oppositional new economic practices are likely to be quickly co-opted, much like organics and the "sharing" economy.

4.4.2. How the "Sharing Economy" Misses the Mark

Some are heralding the benefits of the sharing economy, a new millennial value and economic-exchange system that promotes an ethos of sharing, open-source, and peer-to-peer trade, as revolutionarily casting off the chains of big-business corporatism. However, establishing a sharing economy within the neoliberal paradigm will maintain the problems associated with neoliberalism generally. Capitalism is good at co-opting consumer movements, which is one of the main problems with individual-level boycotts and consumer preference trends that seek to "vote with their dollars." As analysts are beginning to realize, companies like Uber or Lyft are profiting off individuals who trade services via their online platforms, individuals who also receive none of the benefits that they would through private companies. Namely, drivers for ride-sharing apps use their personal vehicles and don't receive compensation for health insurance, fuel, wear and tear, depreciation, interest, taxes, or car insurance (Eckhardt and Bardhi 2015). As characterized by the *Harvard Business Review* "the sharing economy isn't really a 'sharing' economy at all; it's an access economy" (2015, 1). Critics are decrying a commodity exchange process that seeks to dress up old ideas in new clothes. Specifically, ad campaigns for these companies promote a similar ethic as the secularized Protestant ethic explained above. They celebrate stories of workers who chauffeur people around until the moment they give birth and promote lifestyles of sleep deprivation and self-denial of basic needs in order to get the job done (Tolentino 2017). While co-opting the revolutionary ideals of sharing, trade, and mutuality, companies are using these ideals in a capitalist context against individuals, exploiting their work ethics, angst against capitalism, and material needs for sustenance.

While rhizomatic resistance is necessary and important, large-scale change is also necessary in order to maintain the integrity of these growing grassroots movements. Narrowly focusing on consumer movements misses other important avenues for change, such as valuing quality-of-life indicators over GDP, building community relationships, reskilling, and practicing land stewardship. As reflected in my analysis of Asaṅga and el Organopónico in the previous chapters, part of this process is to promote new stories and values that will change our orientations toward people and the planet. From these new stories, we can begin to build and scale up visionary transformative and regenerative socioecological experiments. However, without building this foundation, the corporate co-optation of radical technologies is likely in a capitalist context.

4.5. The Problem of Scaling Up

Consumers are lulled into complacency with the promise of new, more efficient, and environmentally friendly technologies. The truth, however, is that the alluring capitalist myth of consuming more "greener" technologies is not the answer. Much of the mainstream environmental community is concerned with the scalability of sustainable innovations, technologies, and practices.

The term scalability is used in a variety of disciplines to refer to the ability of a system, process, or network to assimilate growing demands and adjust to that growth. It can refer to a business model that has the potential for economic growth or a niche technology that has the potential to grow and be used by more people and/or in more places. Scaling up a specific technology has many social, political, and economic implications. They may range from building new supporting infrastructure, implementing supporting policy, training mechanics and engineers, marketing new values and beliefs, and changing consumption patterns. Ultimately, it's about creating economies of scale, which make production cheaper, and shifting markets to accommodate the mass consumption of a new product, spreading the niche to new consumers and parts of the world (Kemp, Schot, and Hoogma 1998; Spaargaren, Oosterveer, and Loeber 2012).

Context necessarily shapes the outcomes and path dependencies of new technologies. While technologies in themselves may or may not be value neutral, their adoption is shaped by the values and infrastructure of the social, political, and economic context. For instance, there is nothing inherently problematic about the invention of all-electric cars. However, focusing transportation transition efforts away from vehicles with internal combustion engines toward individually owned electric cars rather than mass-transit options, such as high-speed electric trains, reflects a value system that prioritizes individual convenience over larger social welfare. Transitioning to electric cars rather than mass transit requires that individuals can afford them; more resources and space are used for the growing number of car owners; and more waste is produced. Furthermore, the most well-intentioned of technologies in a context that demands growth and expansion whatever the cost will necessarily play by those rules in order to survive—for example, cars are commonly traded in every seven years for the newest model. In this context, each new technology may alleviate a current environmental problem (electric cars' lesser GHG emissions), but will eventually hit another ecological limit (increasing waste from electric cars' toxic spent lithium-ion batteries), creating a new environmental problem (Broom 2019). It is necessary to understand the values and rules implicit in capitalism in order to understand and explain how reforms and technologies are shaped.

Focusing on technological transitions obfuscates the larger, more deeply entrenched problems of capitalism, such as power differentials between the owners of capital and everyone else. These power differentials perpetuate social injustices and environmental exploitation. Extending the electric car example, the Democratic Republic of Congo exports more than half of the world's cobalt, a crucial component of lithium-ion batteries. The mining operations are not only dangerous and toxic but commonly employ child labor, violating multiple human rights (Broom 2019). Designing new technologies will not fundamentally transform power relations that are at the heart of many social and environmental ills. Furthermore, the focus on scaling up niche technologies obscures the fact that many problems, social and ecological, cannot be solved with new technologies alone.

Rather than consuming more of the newest technologies, such as electric cars, creating more resource extraction and waste, consumer societies could change their priorities and values to advance energy and resource-saving lifestyles. These changes could include designing and building walkable and bikeable neighborhoods and communities; building smaller homes that are more insulated or passive conduits of energy; creating local food systems that are more central to the economy; and refashioning a culture that is less obsessed with the newest gadget or fashion trend and more concerned with community. However, many of these solutions aren't compatible with capitalism's goals of growing profits. Walkable communities may necessitate some initial construction, but once built, they require little maintenance and few recurring payments in the way of things like gasoline, car maintenance, or parking fees. Refusing, as a society, to purchase unneeded clothing, goods, or gadgets would devastate the bottom line of marketing firms, consumer goods producers, and media outlets in the business of selling the next new thing. In a capitalist context, these suggestions generate less long-term profit and are thus less economically desirable.

The issue remains that we can continue to develop more "sustainable" niche technologies and create policy that allows them to take over as the primary technologies, but if the structure and rules of capitalism's unyielding exploitation remain, then each technology will fail to be a truly sustainable alternative. As Seyfang and Smith (2007) write:

> Lessons derived from the niche need not be restricted to narrow, technical appraisals of performance. Such 'first-order' learning can be supplemented by 'second-order' learning that generates lessons about the alternative socio-cultural values underpinning the niche and implications for diffusion. (590)

The problem of scaling up, then, is that what must be scaled up are the principles, stories, and values of sustainable innovations and technologies, not necessarily the technologies themselves.

4.6. Beyond Neoliberal Capitalism

Capitalism's underlying values have been, from the outset, competition, greed, and violence, requiring ever-increasing amounts of resources and control over stolen lands. Capitalist values are evident in green-washed movements and technologies—co-opted grassroots movements, environmentalism turned into "green" consumption, and "green" technologies shaped such that they lose their most sustainable attributes. In its neoliberal form, and at its best, capitalism creates lonely, manipulable, and politically meager individuals who consume goods to fill the void. At its worst, it creates impoverished, exploited workers with tenuous employment whose ancestral lands were stolen from beneath them. Surely, we can, and should, expect better from an economic system. This system is not the only way and has been in place for only a tiny fraction of human history (about 600 out of 200,000 years). Reaching a radical sustainability requires that we build a new economic system based on new values and stories. My cases illustrate how we may begin a transition to a more just economic system (for more information about alternative economies see, Daly 1991; Wright 2010; Eisenstein 2011; Schweickart 2011; New Economics Foundation [NEF] 2016).[4]

4.6.1. Making Things Whole: A New Socioecological Paradigm

As Catton and Dunlap (1978) observed about forty years ago, humans see themselves as exempt from the laws of nature. I argue that it is time to re-embed the economy in our socioecological systems. This involves restructuring the three aspects of sustainability, as Flint comments:

> It is notable that "ecology" and "economy" both have the same Greek root, oikos, meaning "house." Ecology is understanding the natural infrastructure supporting the house—the functional dynamics of nature—and economics is the management of the processes of the house—or more particularly manipulating the flow of energy through the house in order to maintain its functional capacity. However, for interplay of ecology and economics to maintain a sustainable flow of energy, there must exist a bedrock of systematic control; this is known as the triple bottom line: ecological integrity, social equity, and economic stability. (2013, 3)

It is nature that supports the household, and our economic system should reflect this reality. Shiva (2016) argues for a pyramid notion of natural, social, and economic systems that positions nature as the foundation of sustainability, then people in the middle, and economics as the tiny point at the top. She contends that our current unsustainable system has flipped the sustainability pyramid on its head, with economics as the foundation. Indeed, economic rationality is often prioritized over social and ecological rationality. Economic rationality is incentivized in the form of the most highly prized object in many cultures: money. Although social—such as family and friends support networks—and ecological—such as ecosystem processes that clean the water and air—incentives exist, they are less often remunerated. Further, cultural norms and values promote competition for money as ethical and just. As mentioned above, the neoliberal paradigm is based on a misunderstanding of Darwin's theory of evolution that emphasizes "survival of the fittest" and competition for resources.

However, cooperation and symbiosis are also necessary for the survival of many species, including our own. To rebalance the scales, we need a socioecological paradigm that not only acknowledges our dependence on and interdependence with nature, but also acknowledges our need for community connections and social equity. Both of my cases offer valuable lessons toward these needs. Cuba and many ecovillages maintain small ecological footprints, close to their fair share of global resources at 1.7 global hectares, while also integrating high human well-being (Litfin 2014; WWF 2014; Global Footprint Network 2015; Sirna 2016). Economic conditions are of course important, but each case demonstrates a community-oriented perspective toward socioecological problems.

The case studies in this book highlight some practices and technologies that are more sustainable because their underlying values are in direct opposition to the values of capitalism. Indeed, some of the most important lessons from the cases in this book are that production and consumption should be *scaling down* and technological designs should be more holistic and work in concert with local ecosystems. More fulfillment should be derived from interpersonal relationships, and communities should work collectively toward self-sufficiency. The innovations from my cases are collaborative, more egalitarian, local, human scale, and labor- rather than energy-intensive modes of production. They prioritize closed-loop systems and self-sufficiency, and they are largely insulated from international markets. These modes of production are simply not profitable in a growth-oriented, global market that thrives on economies of scale with increasingly energy-intensive mechanistic production systems and decreasingly valued manual labor. Thus, scaling up the truly ecologically sustainable aspects of innovative ecovillages and urban farms may not be feasible in a capitalist system. Rather, both cases illustrate what economic conditions could look like when they are not the sole focus of sustainable community and are re-embedded in socioecological systems.

In a radical sustainability, as evident in my case studies, the economy is embedded within socioecological relationships rather than the other way around. The economy is not primary; it simply serves as the metabolic process between households and ecosystems, allowing human labor to navigate the flow of natural resources to sustain individuals and communities as well as maintain ecosystem integrity. A radically sustainable economy is embedded within healthy ecosystems and community life. The economy doesn't define either, but rather it is defined by both social and ecosystems. In a radical sustainability, economic gain and progress could be (re)envisioned to include indicators of increases in quality of life and the health of the biosphere, similar to the Happy Planet Index (NEF 2016). The next chapter focuses on some values that should bolster a radically sustainable economic system.

5

Scaling Up the Values Themselves

Real Utopian Stories for the Climate Apocalypse

> If opposition movements are to do more than burn bright and then burn out, they will need a comprehensive vision for what should emerge in the place of our failing system, as well as serious political strategies for how to achieve those goals.
>
> —Naomi Klein (2014, 9–10)

> The other day I was raking leaves in my garden to make compost and it made me think, This is our work as humans in this time: to build good soil in our gardens, to build good soil culturally and socially, and to create potential for the future. What will endure through almost any kind of change? The regenerative capacity of the earth. We can help create conditions for renewal.
>
> —Robin Kimmerer (as cited in Egan 2020)

An important aspect of social movement work is collectively envisioning an alternative future. Garnering consensus for a unified vision is challenging and is a common source of conflict within movements. Nevertheless, envisioning end goals is a necessary first step toward establishing normative frameworks based on certain values and acceptable means, norms, and behaviors toward those ends (Aptekar 2015). These visions are especially pertinent when considering, as Jared Diamond (2005) concludes in his book *Collapse*, society's survival requires two things, long-term planning and reexamining core values when the old ones no longer serve. Given these considerations, in this chapter I ask, What stories and values will bring about the vision of a radical sustainability? In other words, what stories and values underlie socioecologically regenerative and transformative movements?[1]

Surviving Collapse. Christina Ergas, Oxford University Press. © Oxford University Press 2021.
DOI: 10.1093/oso/9780197544099.003.0006

I also grapple with the limitations of the restorative agricultural cases presented in this book. The Cuban urban farm and US ecovillage demonstrate a paradigm shift that includes community collaboration, holistic systems approaches, nutrients restoration, reconnection with nature, sufficient satisfaction of human needs, traditional skills relearning, reciprocity, and socioecologically embedded economies. However, each case struggles with its own sustainability problems. For analytic purposes, I examine my cases using the three-dimensional definition common in sustainable development literature: social, ecological, and economic. In the following sections, I review literature on Indigenous and traditional Mediterranean food systems to explore past and present potentialities for regenerative agriculture. Then, I identify the innovations particular to my cases in their spatial-temporal context. Finally, I compare my cases to identify stories and values held in common that could be scaled up toward regenerative and transformative, or radically sustainable, food systems.

5.1. Is a Nonexploitative Food System Possible?

In a polemical essay entitled "The Worst Mistake in the History of the Human Race," Diamond (1987) argues that the advent of agriculture ten thousand years ago paved the way for societies based on domination. Through the domestication of plants and animals, agriculture established a lens through which to view the world based on the control and exploitation of nature and peoples. What followed were the beginnings of exploitative divisions of labor, gender inequality, disparate class divisions, and slavery, as well as the exploitation of animals and nature. Thus, the guiding questions for this chapter are as follows:

> Is a nonexploitative food system possible? Are exploitation and domination inevitable? Or is exploitation a matter of degrees? Can you manipulate nature in order to create food, or is that inherently exploitative? Because all food consumption involves killing something, can growing human populations eat ethically? As consumers of plant and animal life, can our relationships to organisms we eat be nondomination oriented? Are exploitation and domination more or less about what we eat or the process by which we obtain food?

Many scholars, including critical plant and animal scholars, acknowledge the need for Western societies to begin listening to nature, as some traditional societies of the past actively did, and still do. It is through the practice of listening and being mindful of nature that we can hear and see the ailments of the natural environment, as well as learn valuable lessons from it (Gaard 2017). Some current forms of agriculture attempt to pay closer attention to ecologies to understand their system dynamics and work with ecosystems rather than against them. Agroecology, as described in chapter 3, and permaculture, as described in chapter 2, both manipulate nature to some degree, finessing synergies and symbiotic relationships. However, the line between domination and manipulation is fuzzy, and cultural narratives and value systems can serve to clarify or further blur this line. These newer forms of agriculture borrow from traditional and Indigenous knowledge sources and utilize similar agricultural forms. Indeed, modern Cubans incorporate traditional Cuban knowledge and agricultural practices in their own version of agroecology (Pfeiffer 2006). In this chapter, I do not mean to romanticize or suggest that traditional knowledges are superior, nor do I intend to perpetuate stereotypical images of "noble savages" or humble peasants.[2] However, there are important lessons to learn from many Indigenous and traditional communities who base their consumption practices around ceremony, sharing, appreciation, and reciprocity. We should be cautioned about applying new technologies, such as agroecology, that appear to have been designed with good intentions, without first understanding more sustainable value systems of the past and present.

5.1.1. An Honorable Harvest

For centuries, Mediterranean fishing communities practiced the *tonnara* as part of cultural and spiritual community celebrations, as well as for subsistence consumption and economic viability. The tonnara is a traditional system of traps for Bluefin tuna. As part of this practice, the fishing community was extremely knowledgeable about ecosystem dynamics, such as the seasons and tuna life cycle, and they would await the return of the tuna to their spawning grounds. During this time, the community celebrated the tuna, including ceremonies and prayers and expressions of gratitude and reciprocity. They carefully ensured that the tuna could reproduce and complete their life cycle. This tradition became a part of Mediterranean communities' cultures and identities for hundreds of years, sustaining community livelihoods and tuna populations, thus demonstrating its resiliency. There are other such examples of sustainable traditions (Longo, Clausen, and Clark 2015).

The Karuk people of what is now northern California think of food in terms of connections and relations. In a global food system that is inattentive to the life it takes and waste it generates, the linear process of production and consumption evokes images of people wrangling and controlling animals and plants without considering the consequences. On the other hand, the Karuks' cultural relationship with the salmon requires that they maintain salmon livelihoods by honoring their life cycle and preserving the ecosystem while also taking salmon lives to sustain their own community (Norgaard, Reed, and Van Horn 2011). They see their relationship with salmon as at minimum bidirectional, dynamic, and changing—a relationship that depends upon give and take. This is a fundamentally different way of looking at life and food consumption than that of the dominant US culture.

Other Pacific Northwest Indigenous ceremonies demonstrate the importance of traditional knowledge by celebrating the life cycle and educating younger generations. For thousands of years, Indigenous communities of the Pacific Northwest have performed elaborate ceremonies during the salmon runs, when salmon make their long journey from ocean environments back to the freshwater ecosystems where they were first spawned. Here they reproduce and complete their life cycle (Longo, Clausen, and Clark 2015). The First Salmon Ceremony takes place when the first salmon make their way upstream. It serves to transfer traditional knowledge, reminding adults and teaching children about the importance of the salmon's life cycle. During the ten-day ceremony, no salmon fishing is allowed, and after fishing for another ten days, all fishing is halted to allow the remaining salmon to finish their journey upstream so that they may reproduce. Values of gratitude, reciprocity, and redistributive sharing are invoked throughout the ceremony.

Kimmerer (2013), a biologist and member of the Potawatomi Nation from the now Great Plains region of the United States, discusses her cultural and spiritual connection to the natural environment. She describes listening to the wind and trees as a means of learning from her "more-than-human" community. In alluring prose, she advances a culture and economy "aligned with life, not stacked against it," an economy that stands in stark contrast and opposition to our current economic system (377). She personifies capitalism as a monster of overconsumption, whose greed and bottomless hunger cannot be sated, because its emptiness stems from a deep loneliness and longing for mutuality. She describes her culture as one of gratitude and regenerative reciprocity, which appreciates the gifts of nature and is responsible for caring for these gifts and giving back. The ceremonies her nation engages in are a way of passing on traditional knowledge about the need for gratitude and reciprocity. In her culture, "flourishing is mutual," and gift giving is a sign of true wealth (382). The antidote, then, to a market system that "artificially creates scarcity by blocking the flow between the source and the consumer" (376) is plenty. She writes that through practices of gratitude and restorative reciprocity we may defeat the monster of overconsumption, which objectifies nature's gifts and harvests for the pursuit of profit while destroying ecologies and leaving invasive species in its tracks. Kimmerer insists that there are ways to harvest nature's bounty honorably.

In *Tending the Wild*, Anderson (2005) demonstrates that Native Americans Indigenous to California had, and still have, a mutually beneficial relationship with nature. Rather than attempt to control or force, they "tend," or care for and look after, nature. They take their role as environmental stewards seriously (358). She argues that Indigenous restorative wildland management has five principles supporting sustainable socioecological relations. First, ecological history and historical Indigenous practices are connected, and they matter for current land management. Second, it is possible and necessary for humans to use natural resources to meet their own needs without destroying the resources themselves. She adds, "Tempered, sustainable use of natural resources demands responsibility, responsibility generates respect for nature, and respect for nature is a basis for sustainable use" (360). Third, transitional proto-agriculture, which combines useful aspects of both gathering-hunting and agricultural societies, provides useful insights for current land management practices. Fourth, sustainable use of resources requires changing culture as well as knowledge production and economic processes. Specifically, creating cultures that value sharing, cooperation, and stewardship is important for creating economies built on sufficiency and restorative reciprocity. Valuing nature and recognizing that humans have a lot to learn from nature are important cultural shifts as well. Fifth, for self-fulfillment, everyone must develop an intimate relationship with nature. We can build a relationship where we grow to understand natural cycles, such as animal migrations and reproductive cycles; where we see our kinship with other creatures and plants; where we creatively express our link to nature; and where we work directly with natural materials to help us connect. In concert, these principles can guide more sustainable and honorable harvests.

Anderson's fourth principle, about changing culture and values, is particularly useful to consider in this chapter. She maintains, as I do here, that the traditions, values, norms, and mores of Indigenous communities, in what is now California, not only reflect an understanding of finite resources, but also affect human behavior and economic relationships toward the land. Anderson describes religious ceremonies, traditions, and norms that foster environmental stewardship:

> The lifeways of California Indians past and present show us that sustainable resource use and human coexistence with plants and animals are based not only on factual knowledge but also on cultural tradition, cultural values, and societal organization. . . . The beliefs and values reinforced by and expressed through these elements of culture define human ways of being with and in nature, and they go hand in hand with the actual individual behavior that is acted out on the land. . . . Knowledge of the finiteness of natural resources must be accompanied by mores and values that limit and constrain resources use. . . . Nondestructive relationships with the natural world need to be reinforced through a variety of expressive and spiritual mediums. . . . Social structure has a strong bearing on how societies relate to the natural world. . . . The wealth-spreading devices operating in many California Indian societies, along with the lack of strong economic hierarchies, discouraged hoarding of resources and encouraged cooperation. . . . The kincentric worldview of California Indians, wherein plants and animals are seen as blood relatives, fostered responsible treatment of other species because it established that they had an equal standing in the world and in fact had much to teach humans. (2005, 361–62)

Anderson highlights how certain values and beliefs about, and relationships with, nature establish more respectful socioecological relations.

As Kimmerer and Anderson demonstrate, nonexploitative food systems are feasible. Resilient traditional food systems can offer lessons about how to live sustainably in place. However, these systems depend on intricate connections between cultures, traditions, place, stories, and values that influence individual as well as collective behavior in relation to nature. Indeed, these accounts suggest that paradigms bolstered by cultural values and stories have material consequences on people and the earth. Informed by these traditions, I discuss the prominent narratives and values underlying my case studies so that we may begin to understand overlapping features and define more sustainable stories.

5.2. Defining a New Story

I hope that we can re-evaluate our current visions and values, conceptualize other possibilities, and tell new stories. Our current global paradigm is unsustainable and unjust, and what is needed is a shift to a new socioecological paradigm. This new paradigm centers the relationship between human and more-than-human nature, and, rather than treating them as separate, treats them as mutually constitutive. In this chapter, I examine new visions and values that will change our orientations toward people and the planet.

5.2.1. A Model of Sustainable Living: Asaṅga

Asaṅga, the ecovillage I researched in the Northwest, maintains a narrative about being the change they seek and modeling sustainability for the surrounding community—and anyone else they can reach (Ergas 2010; Ergas and Clement 2016). Their welcome packet states, "Our home serves not only as a place for us to live, but also as a model of sustainable living for hundreds of guests and tours we host each year." While not perfectly aligned, many ecovillage community activities and villagers' individual practices reflect the values that interviewees expressed to me, as well as those described in welcome materials.[3] Their community organizational structure is positioned in opposition to the dominant capitalist paradigm that promotes individualism and consumerism, preferring to cultivate relationships instead. They specifically seek to live in community, to be stewards of the land, to produce most of what they consume, and to engage minimally with capitalist enterprises. They practice a more holistic version of sustainability than that of mainstreamed sustainable development, one that emphasizes the social and ecological aspects more than the economic. Ecovillagers express various definitions of sustainability that include interrelated conceptions of prioritizing environmental protection, handling conflict in interpersonal relationships, as well as supporting mental health and wellness. While I separate each category—social, ecological, and economic—I only use these distinctions for analytical purposes, because in reality they are interrelated and overlapping. I address some of the overlap as well. What follows are specific practices that ecovillagers engage in that celebrate an oppositional ethos.

5.2.1.1. Social

Ecovillagers value community relationships and see individual and community needs as interdependent. These values are evinced in diverse practices they engage in for community health, integration, and well-being. Specifically, they democratically participate in decision-making, learn and implement communication skills, work on (inter)personal trauma, cultivate personal skills, and participate in community education. Underlying each of these practices are implicit, and often explicit, critiques of a dominant culture that ironically minimizes individual needs and skills in relation to economic gain, undermines social equity by disenfranchising certain groups of people, and lacks education around communication and social cohesion.

5.2.1.1.1. Consensus

Ecovillagers practice a form of direct democracy called consensus decision-making, which I discuss in depth in chapter 2. Consensus requires groups to come to solutions that everyone in the group can agree on. On some more contentious decisions, groups may decide to come to modified consensus when everyone cannot agree on one solution. This is a nonhierarchical form of decision-making that ideally involves everyone in the process, allowing each member's voice to be heard and considering the needs of all group members. In practice, it does not always work out the way everyone would like, but it is an attempt toward social equity and social cohesion. Ecovillagers try to meet every other week, but during the summer not everyone is in town, so they generally meet once a month. These meetings have the added allure of being potlucks where everyone shares food while discussing community matters. During the three meetings I attended, the participation rate was high, with twenty-two of the twenty-seven residents attending one meeting in particular.

One issue caused a bit of tension while I was there. There was community disagreement about a gaggle of geese that was discussed at a few meetings. Carol, a nanny and mother who lived in a geodesic dome, had brought the geese onto the property for her to eat. Many residents did not like the geese because they aggressively hissed and nipped, and they discussed moving them to a local farm. However, an older man, a vegan, began to care for the geese and bonded with them. He told Carol that she could not eat them. He claimed that she had not been caring for them properly, so he'd stepped in and became attached. Carol insisted that the geese were hers to consume. Community members reached consensus during a meeting, when another resident asked if Carol would take a few chickens to eat instead and allow the geese to move to a farm. She agreed, and in the end, the geese were relocated.

5.2.1.1.2. Nonviolent Communication

I participated in a reading group organized by a few twenty-something members of the ecovillage. The group read the book entitled *Nonviolent Communication: A Language of Life* (Rosenberg 2005). The book outlines a form of communication meant to de-escalate the tensest of disagreements by practicing active listening and compassion for the other person and asking a series of questions to discover their underlying unmet emotional needs. When their needs are revealed, the pair can discuss whether or not they can do anything to meet those needs. Participants in this study group discussed the components of nonviolent communication and how they can implement them in their personal relationships. In addition to this reading group, two women on the property who worked as integrative intimacy coaches were trained in nonviolent communication.

5.2.1.1.3. Heart of Now

Heart of Now is a three-day workshop that many community members attend. One community resident facilitates some of the workshops as one of her jobs. The workshops are held multiple times a year at another intentional community. Their purpose is to promote personal growth and healing for individuals. The exercises are intended to help people understand their personal feelings, and by extension, the feelings of others. An intimacy coach explained her work to me:

> I think our culture has this basic agreement that we don't upset anybody's feelings, so we walk through life and we act like we think other people want us to. I see that a lot with my clients. And we get to the point where we don't even know what we want because if we said what we want we either might get rejected or it might hurt someone else or feel like we're invading their boundaries or something like that. . . .
>
> I'm particularly concerned about people breathing and noticing what sensations are in their bodies and tuning into those to discover what ideas are operating in their lives and how they're playing out those ideas

subconsciously. I'm particularly excited about people waking up and creating a life that they want. Not because they're afraid that it's all going to die if they don't but just because they can, like, here we are, be beautiful together.

5.2.1.1.4. Education

Some ecovillagers attempt to educate people in the surrounding community by writing op-eds in the local paper, giving guided tours of the community to students from local schools and universities, and teaching courses and workshops on a variety of community skills. Specifically, when I stayed at the village two residents taught several permaculture courses a year that were open to anyone who could afford them. In addition to these courses, tours of the village usually focused on the natural-building and permaculture designs used throughout. Two women on the property also taught individuals how to practice nonviolent communication. One woman facilitated the Heart of Now workshops, and the other woman called herself an intimacy coach. Huck, an older resident, explained the ecovillage's educational component to me:

> There is kind of an idea of [Asaṅga] being an educational . . . the original part of what [Ralph and Emily] wanted [was] a vision that [Asaṅga] would be educational. We have tours come through here, sometimes thirty, forty people. There's any number [of groups, a research institute], a couple of charter schools, a couple of departments of the university [that] during a course of their events they say, "let's go tour the ecovillage," and so they'll come on in here, and sometimes [Ralph or Emily] or even I will just waltz them through what's going on. Taking that even deeper, is when you come here you are in some state in an educational halfway house, situation. . . . So, there's kind of a social, a strong social suggestion here that you can implement alternatives, and get pretty uppity and strong about it if you feel like it.

5.2.1.2. Environmental

Many ecovillagers understand human societies as interconnected and dependent on nature, and they express valuing nature and environmental stewardship. Some of the ways they express environmental values are through green-building strategies on the property, permaculture design in their garden and community, reuse of waste, and minimal car use.

5.2.1.2.1. Permaculture

Ecovillagers follow a doctrine of permanent agriculture called permaculture. I go into more depth about permaculture at Asaṅga in chapter 2, but to briefly reiterate, permaculture is a form of organic agriculture that takes into account the whole socioecological system, including the surrounding ecosystem as well as the structure of the human community. The housing, community spaces, and garden are intentionally laid out for interaction, collaboration, and mutual nourishment between the human and more-than-human community members. As examples, ecovillagers keep bees and geese to pollinate and fertilize their vegetable garden. In addition, communal tools remain in the woodshop at the center of the village, so villagers can grab shovels or buckets for use in the nearby garden.

5.2.1.2.2. Green Building

The primary property owner, Ralph, is a builder by trade, and he discovered green building in the early 1990s, around the time he bought the ecovillage property. According to the EPA (2016a), "green building is the practice of creating structures and using processes that are environmentally responsible and resource-efficient throughout a building's life cycle from siting to design, construction, operation, maintenance, renovation and deconstruction." Ralph has written about the continuum between green- and natural-building styles. He notes that natural builders "make every effort to avoid using toxic, industrially manufactured materials. They will build houses out of earth, straw and round wood." Green builders are less "radical" in their approach and will use industrial materials that are "environmentally improved," such as low volatile organic compound (VOC) paint. He began implementing natural building at Asaṅga, and has constructed several buildings, including his own house, triplex apartments, and a community center. Two of the structures are made out of cob, a natural-building material made from clay and straw. In addition to being a less toxic building material, cob is naturally fire and earthquake resistant. Some of the other building materials were reclaimed or found from city waste disposal.

5.2.1.2.3. Transportation

Although more than half of the ecovillagers own cars and drive them somewhat regularly, biking, busing, and walking are supported and encouraged. In fact, one of the property owners specifically designed the village so that there is no available parking in the village itself, except for bicycle racks and the driveway to the workshop where his work truck sits. The community welcome packet advocates "the use of ecologically sustainable transportation to decrease the number of motorized vehicles we own and use." In addition, villagers are even given bus passes to promote public transit use. Most ecovillagers engage in a combination of biking, busing, and walking, and driving when necessary.

5.2.1.2.4. Consumerism Critique

Most ecovillagers are critical of consumerism and respond to it in a variety of ways. Some dumpster dive for food and building materials from local markets and manufacturers in order to minimize consumer waste. Some make their own clothes, can their own food, and make their own tools to minimize consumer participation. All but two residents deliberately undertake work that is not tied to material consumption. Some told me that consumerism is a serious threat to the environment. Specifically, Leon, a forty-year-old man from Holland, shared his preferences with me: "I think I felt a strong affinity with living simply. It had a lot of meaning in my life to live simple, consume little. . . . On many levels it sustained me, it was almost like a religion, something to . . . I felt strongly connected to that . . . living simple and being conscious about environmental stuff."

Emily answered my question about what she felt is the most pressing global problem by saying, "For me, probably the single biggest issue is, our house is on fire, the planet. Global warming is real. There's positive feedback loops that will take it out of our control if it hasn't already. And we better wake up fast. And, everything else follows from that."

CHRISTINA: So, hugely environmental?
EMILY: Yes.
CHRISTINA: Do you think this sort of thing relates to other issues?
EMILY: Yes, I do. I think it's all interconnected and so I would say that the reason we go and consume too much at the malls is because we are miserable. And if we just lived happier lives in more connection with people we wouldn't consume as much, nor would we work such long hours which actually hurts the earth by over-producing things. Etcetera, etcetera—so really, it's all connected.

Emily believed that consumption is tied to feeling miserable, working too much, and hurting the planet. This all could be minimized, she prescribed, if we foster connections with people.

5.2.1.2.5. Reusing Would-Be Waste

Ecovillagers reuse as many materials as they can. Each home contains one or two five-gallon buckets where residents put their food scraps for compost. Villagers throw their compost onto a larger compost heap that work traders tend to, which is later applied to the garden. Ralph, the property owner and natural builder, reroutes his household grey water (the cleaner wastewater from sinks, bathtubs, and washing machines) through a filtration system and decorative fountain on the property. Villagers reclaim scrap materials, such as concrete blocks, wood platforms, and metal piping, from nearby manufacturing plants for use in community-building projects. In addition, one villager forages for bread from different markets' dumpsters as well as certain bakeries that give away their old bread.

However, this ethos of saving and reusing comes with challenges. Ecovillagers have trouble throwing some things away even when they are no longer useful. As a result, there are some unsightly pieces of old furniture, disorganized wood planks, and other materials strewn about the main yard. Neighbors have complained about the "junk" and garbage in the village. In response, the villagers built large cob walls around the ecovillage to shield it from the neighbors' sight. A longtime resident, J.T., explains the situation with their neighbors. "I think some of them are curious. Some of them really don't think about [Asaṅga] very often. There was a neighbor, as far as I know his concerns have been taken care of, but he thought that the place looked really messy and he didn't like the place at all so [Ralph] built walls up on the front of the property to accommodate in large part that person's concerns. Well, I hope they're satisfied. It probably still looks like a mess so they probably don't like us, but they can't say anything because [Ralph] bent over backwards to try to accommodate them."

5.2.1.3. Economic

Ecovillagers maintain livelihoods in line with their values. They work as gardeners, nannies, and permaculture teachers. Indeed, many of them work fewer than twenty hours a week, earning just enough to pay for their rent and supplemental food, but not enough for consumer luxuries. Specifically, eight of the twenty-four respondents in my survey worked at least thirty hours a week or more at a job, nine worked part-time, anywhere from one to fifteen hours a week away from the property, while the remaining seven worked solely on the property.

Villagers value their time, interpersonal relationships, and village-related projects more than money or materialism. They grow and glean fruits and vegetables from their community garden and the surrounding neighborhood, or dumpster dive for foods that markets discard because they've reached their expiration dates. Ecovillagers generally work less than forty hours a week, but not out of laziness. In fact, they are quite industrious. They prioritize personal relationships, community-building, and do-it-yourself (DIY) projects, like canning, sewing, building, and growing their own consumables, over conventional nine-to-five employment.

5.2.1.3.1. Cottage Industry

One ecovillager, Ears, spoke of ideas he had to build a cottage industry based at the ecovillage that would help to economically maintain the property. His ideas integrated community values and strengths, and he discussed these ideas with others on the property. He thought it was a matter of time before they began implementing them. In an interview, he shared some of the ideas he considered:

> some of the ideas that I've had, and I've shared them with some people and people have liked them, but it does take a team to organize it. . . . A service of . . . yard transformation. Like for example, if there was someone out there who wanted to convert their backyard into, to regenerate the soil from, we'd have a step-by-step process that we would implement for them in creating a garden in their backyard, because the ecovillage has an abundance of food scraps, it has lots of compost happening and we could be sharing the fruits of our waste almost. With our chickens, we can bring a chicken tractor over there, we can till up, you know, we leave that there for a couple of months rotating it for them every week. Get a couple of loads of leaves delivered to their backyard or front yard or whatever their turning into a garden, mulch it, you know, be adding chicken manure tea, compost tea, all products produced from the ecovillage's waste, pretty much.
>
> And in turn, let's say that household can't make a garden because they are busy at work, they don't have time to make a garden, we can build the garden for them, which is getting them a step ahead. In the process they are paying us the fruits of their labor, the financial energy, which is coming back to us. That's one way that the city and the ecovillage can work in a symbiosis kind of relationship. . . . In the city, I can at least speak for myself,

> it's hard to really start a garden. It takes lots of time and stuff, so we could offer that service. That's one idea for a business we've had. There have been other ideas too. We have considered more theatrical, educational stuff. You know, there's a lot of talented people that are in that business right now, um so there's possibility there. Oh, another one has been setting up . . . a, you know that hay box I built was built completely out of salvaged materials, regular house insulation, aluminum foil, salvaged plywood, um, things that you can find creating some kind of industry from the waste salvaged out of the city, and having some kind of production team here, so where we could build useful items and have an outlet for them somewhere.

Even Ears's economic ideas include environmental logic. He attempted to find a creative way to recycle waste and close the loop of excess from the ecovillage. One idea would be to convert this waste into a profitable small business that would also nourish someone else's garden.

5.2.1.3.2. Movement-Oriented Livelihoods

In their work, villagers care for people and the planet, as all but three of my interviewees had work in line with movement values. During my stay, many of the ecovillagers were self-employed or worked together in gardening, carpentry, landscaping, and natural building. One woman worked as a nanny, another woman an acupuncturist, two women were integrative intimacy coaches trained in nonviolent communication, and there were a couple of permaculture teachers. Significantly, most engaged in paid labor for fewer than thirty hours a week to prioritize other more fulfilling aspects of their lives. The three individuals who worked for local businesses hoped to find alternative employment eventually.

5.2.1.3.3. Work-Trade

Several individuals on the property engage in work-trade. This means that they stay on the property during the day and turn the compost, feed the animals, water the plants, or help the property owners or other residents in other activities. They trade their work hours for a place to sleep. Their sleeping accommodations are not the best on the property by any means. One couple chose to live in a tent near the vegetable garden. They were able to use the outdoor kitchen and bathroom on the property as well as the indoor community center. These residents preferred rustic accommodations to working jobs that failed to fulfill their vision of a sustainable life.

5.2.1.3.4. Reskilling

Reskilling and self-reliance are important values at the community. Community members prefer self-reliance to dependence on corporations to meet their personal needs, and they actively engage many do-it-yourself projects. These consist of activities such as growing their own food, making their own rope, sewing their own clothes, building their own homes, and inventing their own eco-tools, like a solar fruit dryer. Individuals take workshops to learn new skills to further their independence. Workshops included topics like permaculture, green building, and communication.

Amanda, an acupuncturist who engaged in work-trade at the village, explained to me that she made the top she was wearing, and she went on to say, "My friend made the shorts, and I made this necklace. That's a big part of things, learning how to do. Oh, yeah, this kind of goes along with it, specialization. A lot of people . . . are all about specializing, and that's what this world is about, that's how you make more money. Specialize in something. I don't really think that way. I think, actually, I would like to have a broader skill set. That includes making my own clothes, my own shoes, everything."

5.2.1.4. Challenges

5.2.1.4.1. Instrumentality of Animals and Anthropocentrism

As discussed in the consensus process above, tensions existed at the ecovillage between people who ate meat and those who were vegetarian or vegan. While one of the permaculture principles is to care for the earth, some ecovillagers discussed animals and other creatures with regard to their instrumental value, or their value to the human community. However, other ecovillagers had strong affinities toward animals that informed their ethical commitments to vegetarianism and veganism. Whether it was chicken manure fertilizer, beehive honey, rabbit or geese meat, the discussion of other creatures I heard usually centered on what their consumption plans were. I spoke with the twenty-seven-year old single mother, Carol, who raised rabbits and geese for food. "Right now I'm raising rabbits for meat, so they're having babies in about a week [laughs]. And then when the babies are two months old, I'll eat them." She also acknowledged later in the same conversation that this relationship made it easier for her "to pray and talk to the animal." She states, "[I] actually have some hope that the animal will hear me that I'm appreciating the death."

5.2.1.4.2. Political Economy

Private property, landownership, rent, and debt are major obstacles for the ecovillagers. Because of the need to pay rent, a few ecovillagers feel obligated to work jobs that are unfulfilling and counter to their environmental values. One villager confided that she worked in a factory that assembled neon signs for businesses so that she could pay her rent and pay back debt. Aside from environmentally problematic work, property ownership causes tensions and undermines democracy in the village. Because there are three property owners paying a mortgage, financial decisions are ultimately up to them.

Ralph, the natural builder and owner, has said at multiple community meetings that he may need to sell parts of the property. During these meetings, residents have agreed to pay more in rent to help him make the mortgage payments. They also have had to make some concessions, such as eschewing the extensive community interviewing process and allowing some individuals to move onto the property who may not hold the same values or vision. For example, I was aware of one thirty-year old single mother living in an apartment at the triplex with her fourteen-year old daughter. She worked two jobs, one delivering pizzas and the other in collections at a credit union. Most villagers knew very little about her because she was allowed to move onto the property with minimal interviewing. In our interview, she told me that moving to the ecovillage heightened her awareness of environmental issues, which she'd been less concerned about previously.

The case of Asaṅga in the Pacific Northwest United States represents just one attempt at a more sustainable food system and way of life. This approach is situated within a larger neoliberal landscape that affects ecovillagers' pathways toward sustainability. My next case of urban agriculture in Havana, Cuba is situated within a different political-economic context that differently affects their path toward a more sustainable food system.

5.2.2. Sustainable Development in Havana, Cuba

Cuba is as a real-world alternative to globalized capitalism. After the revolution, Cubans prioritized the needs of the populace, but not the environment. Their economy was largely based on trade in sugar monocultures and other cash crops, and they had few subsistence alternatives. Nevertheless, the primary focus of trade was geared toward meeting Cubans' needs over increasing profit margins. As previously discussed in chapter 3, most trade occurred between Cuba and the Soviet Union. When the Soviet Union collapsed in the early 1990s, Cuba became largely disconnected from the global economy. Worse still, the United States imposed trade restrictions that not only prevented trade between Cuba and the United States, but inhibited the docking of cargo ships from other nations as well. Without access to international trade, Cuba confronted economic collapse and, in essence, ecological collapse. Cubans faced malnutrition and the threat of starvation. They had little choice but to make nationwide changes toward food security. To do so, they implemented a number of measures aimed at decentralizing and diversifying food production and moving toward food sovereignty (Koont 2011). The Cuba case exemplifies what relocalization of food production supplemented by global trade could look like if the Global North moved further from consumerism and closer to needs-based consumption. Cuba is exalted the world over for developing nationwide urban agriculture and organic farming, all while confronted by an economic and food crisis.

Cuban culture upholds strong community values that include supporting its citizenry. These values are bolstered by narratives of their socialist revolutionary history—comrades fighting together to oust the US-backed dictatorial Batista regime, eliminating corruption, elevating the poor, and standing tall as David against their Goliath neighbor to the north. These narratives invoke solidarity and values that facilitate collaboration and cooperation toward building a more sustainable nation. The government practices that apply these values include free food rations, healthcare, childcare, and education at all levels. Because food rations were sustained primarily by food imports, rations were cut back some during the Special Period, but jobs began supplying more subsidized or free meals to ensure a minimum of calories (Morgan 2006). Health care and education remained free during the crisis, and Cubans were assured support even when food was scarce. Education complexes became important centers of planning for an agricultural transition and for training the new urban agricultural workers. The Cuban government's investments in its people paid off, because Cubans showed they had the skills and ingenuity to restore food production and the economy. While the Cuba case is not perfectly sustainable, it does illustrate resilience and the complexity of moving toward a more self-sufficient system.

Cuba has implemented urban reforestation projects and organic urban agriculture. Their effort toward sustainability exemplified a flexible top-down regulatory approach that made urban agricultural efforts optional but enticing by including incentives for individuals to develop their own programs. Cubans referred to this as a centralized decentralization (Morgan 2006; Koont 2011). First, the government facilitated Cubans' participation by ensuring their basic needs and supporting their advancement. Then, the government encouraged urban farmers by giving usufruct rights for derelict urban lots to anyone who could make them productive, and further supported them by giving them access to consultants and supply stores. As a result, 60–90 percent of the produce that Habaneros (residents of Havana) consumed now comes directly from urban farms, almost all of it organic. Indeed, many urban farms utilized regenerative agricultural practices, such as permaculture and agroecology. While they sought to incorporate insights from peasant farmers, many of their organic technologies evolved from a larger global agroecology movement (el Organopónico president, personal communication 2010).

Economically, the government incentivized urban farmers by making land rent initially free and now inexpensive and by allowing farmers to take home their profits. The urban farm I worked at distributed profits equitably among its workers on top of their wages, which are based on their years of employment at the farm. Farmers also make a good living in Cuba; their wages are higher than many other types of employment. The Cuban government made these changes out of necessity, and certainly these programs began as top-down and technocratic. However, Cubans' participation and management have been crucial to growing and maintaining the now flourishing projects. At my case site, el Organopónico, cooperative members shaped urban farming practices by engaging in democratic decision-making, experimenting with organic pest controls, and valuing each other's contributions as necessary for the success of the farm.

5.2.3. El Organopónico

I worked at an urban farmers' cooperative to the east of Havana, Cuba. This farm is featured in delegation tours, films, and books on Cuban urban agriculture and is internationally hailed as a model urban agricultural site. The farm occupies twenty-seven acres in the middle of a densely populated neighborhood just south of Cuba's northern coastline. This type of farm is called a Basic Unit of Cooperative Production (UBPC), which means that the 162 cooperative members own the machinery, inputs, and produce, but the government owns the land and charges a subsidized rent. According to my interviewees, the government initially set aside this land for a hospital development. After the Special Period, the hospital plans were put on hold indefinitely, and six founding members, including the farm president at the time of my research, began remediating the land.

5.2.3.1. Social

Aspects of Cuban culture, as well as the culture of the farm, put people before profits. Their revolutionary narrative invokes an image of a small underdog nation of comrades standing in solidarity against their oppressor to the north. The values that stem from this narrative are of caring for their local community and working together for the common good. Cuban practices that reflect these values are laws that ensure residents receive a yearly month-long vacation and a year for maternity leave with reduced pay as well as food rations, free healthcare, and childcare. At the farm, managers assign adjusted hours to caretakers of small children who must walk them to and from daycare or school; allow time off for college courses; and feed employees both breakfast and lunch for free. They hold monthly meetings, of which I attended two, where cooperative members learn about farm business plans and vote on different initiatives.

5.2.3.2. Socialist Revolutionaries

Billboards, murals, and commercials in Cuba serve as daily reminders of Cuba's 1959 revolutionary victory. At the farm, I heard members of the older generation who lived through the revolution express gratitude for the changes that occurred afterward. However, Cubans who came of age during or after the Special Period of the early 1990s express a longing for more opportunity and material possessions. One older man at the farm discussed his gratitude for the revolution with me:

> Look, I'm grateful. Why? A *guajirito* [farm boy] from Sierra Maestra, in Sierra Maestra the rebels, Fidel's people came together and fought against tyranny. That's why they call it Sierra Maestra [Master Mountains]. It's in Cuba, in Santiago de Cuba. Then, I was a guajirito that was born in the country. The revolution gave me the opportunity to study; I studied and became an engineer. The revolution made me an engineer; I married an engineer. She [the revolution] gave me a house, even though I paid for it, she [the revolution] gave it to me. She [the revolution] gave me a house, my wife, me, my two daughters, a scientist and another. She [the revolution]

> gave me two grandchildren. Do you understand? I have the revolution to thank. Although there are things that, aware of them or not, are mistakes of the revolution, but sometimes things, people are more fixated on the bad than on the good. Here, you get seventeen years or twenty years, or thirty years, or forty years of studying and are not charged a penny. If you go to a hospital and you are in a hospital for twenty years, there at the hospital, they do not charge you a penny. When you retire, the state assures you a retirement, not very large but it assures you something until you die. Before the revolution, people went hungry, you know what I mean? After the revolution, no. Before the revolution, the only ones who bought meat were those who had money; the poor did not eat meat. In this revolution, bad or good, you eat something.

In this same conversation, he concedes that current economic conditions in Cuba are grim. He continues:

> Of course, here in this country we made 200 to 250 thousand houses a year when we were good with the Soviet Union. Since that fell, now 100 thousand would be a lot. You understand me? Why? Because the most serious problem that Cuba has in the whole world, is the problem of housing, of living. Here we get married and we always live with our whole family. But now that mentality is changing. And everyone wants to get married and have a house. Like you there [in the United States], you marry and live independently from your dad and mom. Here, more or less, people want to do that too, but it's difficult. You know what I mean? Here, everyone would like to have a car but cannot because there is no oil. You understand me? So, there are things that make you say "But, I'm an engineer, I'm a lawyer, I'm a doctor, I'm a scientist." And so, that's one of the problems, the economy is very bad. We, the only thing that we have here is tourism; that is the one that is helping us now. Last year, two million, five hundred thousand tourists came to Cuba, and that helped us a lot, because it brings money, and with that money, we go outside and buy medicines, buy food, buy all kinds of things.

Because of these changing cultural conditions and the state of the economy, every younger Cuban I met told me that they wanted to move to the United States, and, since I left Cuba, at least one of them has moved.

5.2.3.3. Compañeros

Ironically, Cuba's isolation from the global economy likely buffered them from some of the more socially isolating consumer technologies. Some socially isolating features of US society include smart phones, ubiquitous internet access, lack of free public spaces, large single-family properties, and limited connections or competition between neighbors. While I was in Cuba, only certain workers, like researchers, had internet access at work, and most people had little to no way to connect to the internet (though this has since changed).

Shielded from some of the isolating features of consumptive lifestyles, Cubans communicate with each other daily. People know their neighbors. The neighborhood I stayed in was entirely made up of high-density apartments where people regularly sat outside in their stairways talking to one another. Many Cubans must live with their extended families due to a serious lack of housing; thus, privacy is limited.

Compañero can mean comrade or colleague and is a common way that Cubans address each other, even strangers. They maintain a culture of talking to people in their neighborhoods, including people they do not know. Most people walk and take the bus to markets and to work, instead of driving alone, as residents of the United States and other industrialized nations do. In the streets, they chat, play music, dance, and sing.

I interviewed three sisters who worked together at the farm, and they discussed the importance of community: One twin said: "Focusing on the collective is very important. It's bad to be like most people in the world who only care about themselves. Your neighbors, family, and friends are important. Everyone should work together toward the collective good." Her twin chimed in, "Just going home and being isolated in your apartment in front of the TV isn't good for anyone. People need each other." Their older sister added, "There is a Cuban saying, 'your neighbors close by are your family.' Cubans live more collectively. You go say 'hi' to your neighbors, ask how they are, borrow some sugar, drink a coffee, sit and talk." The sisters continued their discussion of the importance of community between themselves.

Some people disagree that all Cubans have their neighbors' best interests at heart. In a conversation with an agronomist, she told me a story about a neighbor she believed to be un-neighborly. She claimed that this neighbor only called when she needed something and never returned the favor.

5.2.3.4. Work Meetings and Decision-Making

At the urban farm, cooperative members meet monthly to discuss future directions of the farm. The farm maintains a relatively hierarchical organizational structure, with a farm president and bosses for each sector. During these meetings, the sector bosses bring new matters to the attention of the cooperative. However, they maintain democracy through voting, and every cooperative member votes on each matter. Among issues they voted on while I was there were whether to hire new members, if they should sell produce in local markets, and what plants to produce the following year.

5.2.3.5. Environmental

Cuba's environmentalism largely came out of a dearth of natural resources and capital after the Special Period and in conjunction with the ongoing US embargo. However, because Cuba had invested in its people by educating and feeding them and maintaining their health and well-being, they have an abundance of human resources. These include skilled doctors, scientists, engineers, and laborers with the ingenuity and creativity to solve the problems associated with their crises. Cubans boast that with only 2 percent of the Latin American population, they produce 11 percent of Latin American scientists (Morgan 2006). While they must ration resources to ensure that the food and machinery are evenly distributed and not hoarded among the wealthy, the rationing system comes out of resource scarcity and cultural values that prioritize equitable distribution among the population, rather than an ethos of resource conservation. Regardless, their values of social equity are met most efficiently through working with nature in ecological practices like agroecology. Thus, environmental sustainability may be necessary to ensure the most efficient distribution of resources to the populace.

5.2.3.6. Structural versus Individual Environmentalism

Cuban farmers are not all "environmentalists"; some throw garbage on the ground and use animals instrumentally, and some lack explicit environmental ethics. I witnessed two coworkers throw food wrappers on the ground outside of the farm, and when I worked in the agribusiness sector filling spice packets, I was instructed to throw plastic trash in the bushes. When I looked into the bushes, they were full of plastic garbage. Garbage, such as beer cans, plastic cups, and cigarette packs, also line the alleys and walkways around the farm. I could see that this practice was common in high-traffic throughways off the farm as well. Farmers instrumentally use animals by having oxen plow fields, keeping caged rabbits, goats, and birds on the farm, killing chickens and pigs for food, and shooing away starving dogs that live on the farm.

In my interviews, I asked my respondents what sustainability means to them and how they felt about the natural environment. Four out of thirteen respondents expressed an explicit environmental ethos. The former doctor who changed careers and now examined insects and worked on biological pest controls, told me what she liked about her job:

> I like to do everything. I even like cleaning the floor, because, I don't know. Everything I do, I establish an exchange with every little thing, with every detail, and, so, we are well . . . and I like to be exchanging all the time with things, with the plants, with the animals. I like to plant plants because I see them grow, I see that I can achieve things with things as simple, as nice as a plant, to see them alive, to water them for them to grow, to feed them, to make them pretty. They will then help me raise my animals, see? It is an important exchange, and insects similarly. With the insects, to give them food, watch them grow and help nature a little to express itself through me.

She sees each relationship as valuable and connected, even the relationships between the plants and insects.

Eight of thirteen interviewees discussed the instrumentality, or importance, of nonpolluting activities for human health. They focused on environmental health as important but secondary to human health, and suggested we preserve the environment to the extent that we maintain human health. A woman who worked in the agro-industry processing area commented, "They try to help nature but not everyone does it, do you understand me? Not all of them do it. There are people who do not understand that by helping nature they are helping themselves." Similarly, an older man spoke of the importance of nature:

> Because we all know that man has to live with the environment. If you harm the environment, man is also harmed. If you develop a series of elements that benefit nature, man also benefits. If man does the opposite, what happens? The environment is affected either by our use of water, either by the use of the land, or by the use of trees. All that must be protected, one way or another we must protect it, because if not, the population is affected in the long run, everything, I say, the population of the whole world because

these big fires are happening, these big downpours, these great cyclones, they bring consequences that affect us.

While individuals do not generally express explicit environmental concern, their social structure supports an environmentally sound material reality for its people. Most Cubans engage in very little luxury shopping, because they lack access to many luxury items, so consumerism is not a driver of environmental degradation in the country. Cultural values based on equality and the equitable distribution of resources support sufficient distribution, ensuring everyone receives enough to survive. Their food rationing system is based on the UN minimum calories per person per month, and before the crisis citizens received a full four weeks of rations. However, as food imports decreased significantly after the Special Period, rations were reduced over time (Morgan 2006). The US embargo has affected Cuban imports as well and, as an island, resources have always been limited. Thus, many people conserve materials out of necessity rather than because of environmental values. For example, Iset, the woman I stayed with, wanted to buy tomato paste from the farm, but she didn't have an empty container and had to wait until her family finished drinking soda from a plastic bottle. After the soda was gone, she washed the bottle and took it to the farm to be filled with tomato paste. Many younger Cubans wish to consume more clothes, cars, food, and entertainment. However, Cuba's environmentally sustainable reality is unaffected by some individuals' desire to consume more or the lack of an environmental ethos. Cuba came to environmental sustainability because they lack resources and distribute limited resources to as many people as possible. Working with the environment, rather than imposing will upon it, proved the most efficient means to support themselves.

5.2.3.7. Agroecology

Agroecology is a technique that takes advantage of ecological synergisms by using biodiversity, like integrating animals into rotational grazing systems with crops, diversifying with polycultures, and employing insects' predator/prey relationships for biological pest control. The farmer works with nature to allow ecology to do much of the work. While some of the techniques they use come from traditional Cuban knowledge and practices, the president of this UBPC informed me that they learned many of their techniques from University of California scientists.

On site, the farm consists of eleven sectors, or workstations, that each work in relation to the others. Specifically, the farm has a livestock area where bulls are employed for plowing, and their manure is saved to feed the worms at the vermiculture station. At the vermiculture station, worms, compost, manure, and rice shells are cared for. Worms living in large, shaded vats consume the manure and expel their own waste or humus. The humus, compost, and rice shells are combined to make soil and fertilizer for the seedlings. When the soil is ready, it is taken to the plant nursery where some of it is mixed into trays along with seeds, organic pesticides, and water. The seedlings stay in the nursery until they grow large enough to be planted in the fields. Once out in the fields, they are tended to by farmers and scientists who manage pests.

There are two scientists on staff, including a doctor, that work in the rustic laboratory observing insect interactions and preparing organic pesticides. They manage pests in several ways. Plants like marigolds and sunflowers are strategically planted to attract or repel pests away from food crops. The scientists also manage insect populations by introducing natural predators and by making and applying organic pesticides. However, the farm's doctor, who is also an insect specialist, told me that she does not consider any insect a pest because they all serve a function in nature. Some are just more beneficial for the plants than others, and she cultivates the beneficial bugs to release them at times when there are more plagues. The scientists also experiment with other forms of pest control. The doctor tells me she is observing a certain kind of bacteria that can kill the larvae of certain insects.

As the crops in the field mature, they are harvested and sent to one of three stations. If crops are too ripe at harvest, they may be sent to compost or the agro-industry area where they are processed into tomato paste, garlic paste, spice packets, or pickled vegetables that are then sold in the vending area. The most attractive produce is sent straight to the farm's street-front vending area to be sold to people in the community. High-density apartments sit just across the main avenue from the farm, and people in the neighborhood stop by to purchase produce, spice packets, sugarcane juice, or ornamental plants. The vendors collect the money, which the economic sector in the farm's front offices then counts and processes. These offices are also where administrators manage workers' salaries and conduct the sales of bulk produce to Havana restaurants and hotels. Profits pad workers' salaries and are reinvested in the farm.

5.2.3.8. Reusing Waste

In their own mechanic shop, farm workers often use parts from cars or broken machines in order to fix machines still in use. People at the farm collect used glass bottles and clean them for reuse in the agro-industry section. Sometimes they have trouble finding enough used bottles, and community members will bring in their own bottles from home, as Iset did, to be filled with tomato paste or other processed condiments. Many Cubans I interacted with reclaimed and repurposed various objects because of the lack of resources. Whatever the motive, repurposing glass bottles to sell food is a worthy endeavor. And of course, the farm reuses plant and animal waste for compost and fertilizer to enhance the soil.

5.2.4. Interdependence and Gratitude

Cuban urban farmers appreciated the social, economic, and environmental benefits of urban agriculture. I asked a student of agronomy what urban agriculture meant to her. She acknowledged the interconnectedness of the different aspects of, and said she was grateful for, urban agriculture:

> It's an alternative to hauling food long distances. It's a job for people who are unemployed. It cleans the air, atmosphere, and environment in the urban centers where there is a lot of movement and pollution and where the contaminants collect. It brings food to people. In the peripheral urban areas where there is land, why not use it for food?

5.2.4.1. Observing and Listening to Nature

A gender analysis had been conducted on the urban farm prior to my visit. The analysis found that the women at the farm were the first to notice pests on plants because they took the time to look at them. The doctor who became an insect scientist monitored insects and made sure that enough predator populations existed to keep pest populations in check.

The woman studying to be an agronomist explained how cultural gender relations work to establish these patterns:

> The man is much more, he goes through walking fast, there was, I do not know, a pest or a disease, and because he is walking quickly because he is

> working, he does not see it. The woman is more observant, more dedicated. They may be in the same space, but the woman passes and what the man did not see, the woman saw it. And then you can say to the man "look, you have a pest, or there is a planter with problems," and she can alert the man. And it's because it's a cultural problem because many times the country man, the man who works the land itself is sexist. The woman is for cooking, for the house, to raise children. He cannot imagine the woman corrupting agricultural work. But hey, here you see that it does not happen like that. Here the woman is, she is also an urban agricultural worker.

She acknowledged the importance of women's attentiveness to plants. Paying attention to their needs is crucial to the functioning of organic agriculture.

5.2.4.2. Economic

Soon after the Special Period, Cuban farmers worked to produce free or cheap food for local hospitals and schools. In recent years the need has lessened, so farmers can focus more on profitable exchanges, such as the ones in their neighborhoods, local markets, restaurants, and hotels. Only one of my interviewees indicated that she would prefer to do some other type of work. Almost all told me they enjoyed this work for a variety of reasons. Specifically, they appreciated the pay, the location close to many of their homes, the reward of being able to see the literal fruits of their labor, and the fact that they are helping people.

5.2.4.3. Livelihood

By Cuban standards, the workers on this farm earn a decent wage, even higher than most government jobs pay. During my stay at el Organopónico, wages were capped at 1,000 Cuban pesos a month and were calculated based on number of years employed at the farm. Workers begin at 400 pesos a month and earn a raise every five years. When sales exceed overhead costs and payroll, profits are distributed among workers or reinvested in the farm. Cooperative members are able to keep costs down, and wages relatively high because the farm engages in all production and sales activity on site. They do not have to pay middlemen or other businesses for food processing or transport. Further, they use minimal farming inputs, and most inputs are produced from resources already at the farm, such as compost and manure. Since many workers live in the surrounding neighborhood, they can walk to work; hence, transportation costs are minimized for them as well. Beyond these practical advantages, most of my interviewees described this work as beautiful and fulfilling. One interviewee, who came out of retirement to work at the farm, told me she worked there because:

> The first characteristic was that I am close to my house, the proximity to my house. I live three blocks from work. . . . It is a very healthy job, a healthy job since I am in an environment, in a healthy environment of trees, in the open air, and in addition, it is well paid, well paid, and well paid. The collective is very good, I feel I'm in a good collective.

Other farmers recognize the good they are doing for the community, and they feel good about helping others. Two sisters told me that they like their work because they get to see the fruits of their labor and "people need your work, and you know that you are helping others." Of course, the Cuban economy is still hurting, and the urban farm faces many real day-to-day challenges.

5.2.5. Challenges

5.2.5.1. Sexism

Entrenched cultural gender norms distribute and divide labor along gender lines. Women perform tasks that require less movement and are in shaded areas, such as sitting under thatched roofs planting seeds or working in the main offices counting money or keeping track of workers. Men are generally more active and often work in the sun, out in the fields planting, harvesting, or pushing wheelbarrows full of soil. Men and women alike uphold these divisions by naturalizing them and expressing resentment toward them at the same time. Women also resent having to do housework when they get home from work, and some men who work in the fields resent that women seem to have easier jobs at the farm. I discuss these relations in more depth in chapter 3.

5.2.5.2. Anthropocentrism

As previously noted above, farmers tend to discuss the environment, climate, plants, and animals in terms of how they affect people, rather than the value they have in their own right. These instrumental narratives about animals and plants affects relationships that people have with them. Specifically, one agronomist noted that men are less attentive toward the plants and are less likely to notice diseases or pests. Cubans' attitudes toward the environment are changing, however. An accountant admitted that she had little environmental awareness until she began working at the farm. She also noticed more because the government had begun delivering public service announcements about the environment. When I asked what she thought about the natural environment, she said:

> Well, I have more ideas about the environment since I've been working here, because I'm surrounded by nature, apart from my husband. He is a nature lover and he, please, has taught me many things really. In addition, on television, Fidel himself, talks about it all the time, so, right now he is imparting some knowledge about the fact that nature really matters in the environment. But before in our culture it was not . . . yes, you went to school, they gave you some more or less, but everything was there. But the media are disseminating a lot of propaganda about this [the environment]. That is, it is not so much work or the place where you work, but the media that are emphasizing this [the environment]. But it is also not true that all people have ideas or accentuate this, no. Generally, he who works directly with the earth, is he who has more ideas about things as they are, because people do not consider this [the environment] important.

5.2.5.3. Rationing

The government's rationing and distribution system goes beyond food rations, as many resources, including machines, are scarce. At the farm, if they need new machinery, they have to request approval from the government before they can buy the item. Awaiting approval causes problems on the farm because it can halt production and make their work less profitable. When I asked workers about obstacles on the job, some complained about bad weather, but most interviewees complained about a lack of resources:

> Obstacles? When we want to do something, and we are not decision-makers. For example, even though we have money and want to buy an ice machine for the sugarcane juice, we cannot make the decision to buy the machine. We have to ask the government for authorization to sell us the machine, and they decide if we need it or not. And many times, when they finally give us permission, yes, we can buy it, they have delayed us so much in paperwork, bureaucracy, and permission that when we get to the place where they sell the machine, they ran out, there are none.

5.2.5.4. Usufruct

Usufruct rights exist when a property owner allows someone else to use their land barring its destruction. The urban farmers at the farm I worked at were never in any danger of losing their land. However, the government does still own that land, and the farms do pay a subsidized rent to keep the land productive. In theory, if the land became more valuable if put to another use, the government has the final say about its distribution.

5.2.5.5. Convincing Young People to Farm

The farm president informed me that younger people in Cuba do not want to farm. Cuba has an ongoing educational effort to change people's minds about farming, because they need the labor in order to feed and sustain the populace. The education campaigns include children's school groups, advertisements on television and billboards, and educational tours at the farms. I explore this in greater detail in chapter 3.

5.2.6. Holistic Lessons from the Field

Both ecovillagers and Cuban urban farmers provide useful insights for developing a radical sustainability, one that is at once socially and ecologically transformative and regenerative. They integrate more holistic systems thinking into their projects, re-embedding economic relations in community and ecology. They also share some core values and practices, even if they arrive at them out of different circumstances and needs. Both sites reuse urban waste for their own do-it-yourself projects. In Asanga, villagers reclaim city waste to repair broken items or to make new, eco-friendly items and structures. At el Organopónico, farmers gather empty glass bottles from the surrounding neighborhood, clean them, and fill them with garlic paste, which they sell at the street vending location. In the farm's repair shop, the mechanics reuse scraps from broken machinery to fix machines still in use.

In terms of agriculture, both sites seek to work with nature, to integrate regenerative practices, and to value reciprocity. At Asanga, residents use a technique called permaculture, while farmers at el Organopónico employ agroecology technologies. I also met urban farmers in other parts of Havana who integrate permaculture practices on their land. Both styles of farming facilitate ecological synergism, reintegrate organic waste, employ polycultures, and apply biological pest controls. In each style of agriculture, care for the earth is invoked explicitly, by returning nutrients to the soil to maintain its health and the diversity of microorganism life.

In addition, both sites value cooperation to reach common goals and acknowledge their interdependence with each other and the natural environment. They each engage some direct form of democratic decision-making during regular meetings, where the associated producers work together to reach common goals. El Organopónico is a cooperative whose members own farm resources in common and make decisions about these resources communally. In addition, each part of the farm is dependent on every other for its success. Thus, each sector must work to fulfill the needs of the next. The plant nursery relies on the compost and worm humus from the vermiculture sector, and the farmers in the field rely on the starter plants to plant in the field. At Asaṅga, villagers participate in regular meetings to make group decisions, residents offer to pay more rent when Ralph has trouble making the mortgage, and people gather regularly for work parties to beautify and restore the property.

People at both sites value education and the transfer of information. They prioritize education in a variety of ways. In Cuba, higher education is free to its populace. The woman I stayed with was finishing her degree in agronomy and was given time off from work at the farm to attend classes. Farm training and consultants are also available to those in need. The farm site itself is an educational center and gives tours for international sustainability delegations. Further, the farm welcomes researchers, documentarians, and nongovernmental organizations that seek to disseminate their powerful lessons. The ecovillagers also welcome tours from local schools and community groups. They allow researchers and film crews to document their experiment in social and environmental relations for the benefit of others. In addition, ecovillagers teach courses in different aspects of sustainability, such as permaculture, nonviolent communication, and personal well-being. Ecovillagers charge fees for these courses to help support themselves but accept sliding scales and work-trade in lieu of money from those who do not have the means to pay full fees.

Economic relations are different for Cuban urban farmers and ecovillagers in the United States. The Cuban economy is a planned economy designed to more evenly distribute resources among the population by setting low prices, rationing resources, and supplying free education and healthcare. Urban farmers at the cooperative farm the city for a living. It is a form of employment that offers them a comfortable lifestyle in Cuba. They are able to live off of their farm wages because they pay a subsidized rent for their land and are vertically integrated, meaning that they manage all stages of food production and have no middlemen or corporations they must negotiate with. Wages are distributed based on duration of employment and are capped, keeping farm costs down. In times of surplus, profits are distributed evenly. They make group decisions about wages and equipment purchases, so everyone's needs are heard and honored. While making a living is important to everyone on the farm, so is the personal satisfaction of doing work where they literally see the fruits of their labor and feed the local community.

Asaṅga's economic conditions are very different. They are situated in the United States but are living in opposition to its economic system. Some ecovillagers do in fact express an explicit critique of capitalism, though others do not. However, most critique many aspects of US culture that serve to bolster capitalist relations and harm the environment, including consumerism. Ecovillagers must pay rent and/or a mortgage on the land they live on; they must work in order to pay said rent; and a few of them must work jobs that do not align with their values in order to earn that money. Urban zoning laws in this city prohibit the sale of produce grown there; therefore, they cannot sell the produce they grow on their property. Some enterprising ecovillagers find ways to make money from skills or talents less valued in the larger culture, such as teaching permaculture courses. Most ecovillagers seek to work fewer hours in wage labor to foster building interpersonal relationships and other subsistence activities. None of the ecovillagers spend money at shopping malls or similar outlets for luxury items, although some do engage in thrift store shopping.

I would like to address a common concern of environmental advocates and scholars. They often ask, what can be scaled up from these innovative sustainable technologies so that people all over the country or the world can use them? While I believe it is important to scale up many practices from both Asaṅga and el Organopónico, I want to caution against the piecemeal application of regenerative farming that doesn't consider community or social justice, or applying direct democracy without considering environmental cycles or ecosystem dynamics. A radical, holistic vision of sustainable community, one that bolsters both the social and environmental aspects, requires *systemic* change that is at once transformative and regenerative.

5.3. Bringing the Scales Back Together

Some analysts offer prescriptive paths toward sustainability that prioritize a specific scale or type of human activity, such as the individual or local community and local economies. In other cases, scholars and activists prioritize large-scale national or global change, as in planned economies and global regulations. However, scholars of social movements have useful information regarding successful movements of the past (Goodwin and Jasper 2015). The strategies and tactics must be diverse and must occur at every societal scale: binding global climate treatise, national environmental regulation, shifting economic priorities, relocalization of trade, and changes in individual behaviors. Just as ecosystem diversity is more resilient to shocks to the system, capitalism has resilience because its tentacles are far reaching and supported by corporate enterprises, consumer culture and ideologies, policies and regulations, and international organizations and development programs, as well as individual investments. Thus, socioecological justice movement strategies must replicate this diversity to target beliefs and values, narratives, laws and regulations, divestment from corporations, complete disengagement in consumer culture, as well as confrontational tactics that serve to halt production processes. The political-economic opportunity structure literature may offer some insights as well. Looking for cleavages among elites or toward sympathetic bureaucrats may offer some opportunities for change (Pellow 2007). In some cultures, the strategy may resemble authoritarian-planned economic regimes; in other societies the tactics may be more grassroots in form. Large-scale change is needed, and grassroots movements are important, but they are not the only players in biospheric change.

One of the goals of modernization and capitalist development is to create consumers out of people who once subsisted off common land and resources. As development agencies and multinational corporations attempt to privatize and commodify everything, such as water and land, former peasants are turned into laborers and are forced to work to pay for what they once freely derived directly from nature. In many cases rural populations are forced to migrate to large cities where, as the generations pass, they lose traditional knowledge systems and means of self-sufficiency, ultimately becoming dependent on capitalist enterprises for their food, clothing, and shelter.

However, food sovereignty, do-it-yourself, and back-to-the-land movements are all groups of people seeking to regain that traditional knowledge, to become once again independent from private institutions. Both urban agriculture in Havana, Cuba, and the ecovillage in the Pacific Northwest have disengaged from global capitalism to varying degrees and at different scales. Cuba as a nation was largely economically isolated as a result of US embargo. The ecovillagers, as a grassroots movement, attempt to produce and upcycle, or reuse, most of the items they consume. While neither is fully detached from global capitalism, this approach, if done on a large enough scale, could serve to undermine the growth mechanisms of capitalism. This may seem like stepping backward from "progress," but I argue that most of us would be freer if we disengaged in and divested from capitalist enterprises. Relying on the market to meet our basic needs keeps people in jobs they hate, doing work they find unethical, unfulfilling, and boring just to keep food on the table and a roof over their heads. DIY movements remind us that we have all the tools we need, as well as our community assets, personal abilities, and ecosystems, to sustain us. From there we can forge new paths not marked by keeping up with the Kardashians (or Joneses, as they once were). The cases I have presented here illustrate potential pathways toward independence from capitalist enterprises and toward regenerative cultures where growth in community and biodiversity will be valued above growth in capital.

5.4. Beyond Sustainability and Resilience toward Radical Sustainability

In a book aptly titled *Questioning Collapse*, a group of archeologists and anthropologists critiques Diamond's *Collapse*. They argue that the history of human societies is not one of recurring collapses, but rather of resilience and survival (McAnany and Yoffee 2010). Even when communities must relocate or disband certain practices, they overwhelmingly do survive, even if in smaller numbers or at different locations. The book's message is a hopeful one, one of resilience. My message here is similarly about hope and recovery.

However, the above cases demonstrate that we must move beyond the narratives of resilience and sustainability toward social and ecological regeneration and transformation, or radical sustainability. The EPA (2016b) definition of sustainability is "to create and maintain the conditions under which humans and nature can exist in productive harmony to support present and future generations." This definition begs the questions of sustaining what and for whom? Are we sustaining lifestyles of excess for a minority of economically privileged individuals at the expense of others or nonhuman animals? The current trajectory of sustainable development would suggest that this is indeed the case. Resilience suffers a similar definitional problem when it is defined as "the ability of a system to absorb disturbance and still retain its basic function and structure" (McAnany and Yoffee 2010, 10). This notion of resilience still fails to elucidate what the system is that we are trying to maintain, or question whether maintaining the current system should be the goal at all. The tools already exist to create a new system built on a transformative politics that values direct democracy and listening to the needs of all earths' beings, while simultaneously enhancing the regenerative qualities of life. Thus, if resilience is about the ability of a system to absorb disruption, and sustainability is about maintaining the conditions necessary to support future generations, then I argue that neither of these go far enough. Rather than maintaining, we should be doing the work of making things better and leaving the earth in a healthier condition than we entered it.

A radical sustainability framework then is holistic with socioecological interdependence at its base, followed by equitable regenerative relationships, and facilitated by an economy that bolsters these interdependent, equitable and regenerative relationships. A radical sustainability is necessarily anti-capitalist—as capitalism is a system that organizes based on inequities—and anti-authoritarian—as direct democracy establishes untethered feedback between individuals, communities, and ecosystems potentially experiencing harm. A radical sustainability is community oriented (yet celebrates individual and cultural difference), encouraging self-sufficient communities that are premised on ecological principles. Radical sustainability is not premised on self-denial or obligation; rather, it is based on pleasure and reciprocity (Brown 2019). A radical sustainability also is relational and based on ethics of caring, sharing, and sufficiency, consuming only what one needs and sharing the rest. Further, a radical sustainability is regenerative, giving back to the community and the earth more than what was received. A radical sustainability also appreciates diversity in all its myriad forms. Finally, a radical sustainability attends to the local, both community and ecosystem, because localized traditional and evolving knowledges are more attuned to socially and ecologically efficient practices and can recognize problems as they manifest; thus, they are more adaptive and able to change. To more tangibly exhibit a radically sustainable framework, I turn to the Indigenous and traditional societies as well as the cases presented in this chapter, as they illustrate equitable and regenerative relationships.

Much like Asańga and el Organopónico demonstrate—and similar to the spiritual and ceremonial fishing and harvesting narratives of the cultures described above, the Pacific Northwest Native Americans' First Salmon

Ceremony and the Mediterranean communities who practice the tonnara—we need a regenerative and transformative framework, or a radical sustainability. The traditional cultures described above speak of plants and animals relationally as family, as part of their own lifeblood.[4] This view changes the stories and general orientation that community members have toward other creatures. Rather than justifying harm by othering or devaluing them or objectifying them as food, these traditional cultures treat other creatures with respect, or the way they would treat family members. And they go further by understanding and promoting the prosperity of the surrounding ecosystem so that the fish or other species they consume will thrive. They speak of this as a reciprocal act and express gratitude; creatures give their lives for the community to eat and survive, and, in return, the community ensures that their habitat and species flourish. They surpass maintaining and surviving by building burgeoning and healthy ecosystems. By bolstering the needs of some of the most vulnerable in their eco-community, the fish, everyone in the community thrives.

While the cases of Cuban urban agriculture and the ecovillage focus less on spirituality and ceremony, they do further illustrate some values that we could scale up toward regeneration. These are the values of reconnecting and relating to the plant and animal world. They do this by working cooperatively with nature and community; appreciating and reinforcing diversity in all its forms; listening, observing, and considering human and natural communities' needs; taking only what they need or what is sufficient; learning life-supporting skills, such as regenerative agriculture; supporting and helping others; practicing mindfulness; sharing knowledge and abundance; recognizing interdependence with people and nature; giving back in acts of kindness and reciprocity; and expressing our gratitude. As communities strive to build new or alternative economies, these are among the values they should build on, values that prioritize relationships with socioecological communities rather than individual economic gain. These are the values of a radical sustainability.

Conclusion

There Is No Future That Is Not Built in the Present

> The most anti-capitalist protest is to care for another and to care for yourself. To take on the historically feminized and therefore invisible practice of nursing, nurturing, caring. To take seriously each other's vulnerability and fragility and precarity, and to support it, honor it, empower it. To protect each other, to enact and practice community. A radical kinship, an interdependent sociality, a politics of care.
>
> —Johanna Hedva (2019)

On a cloudless sunny afternoon in the summer of 2019, about a decade after my last visit, I returned to Asaṅga. A pleasant breeze relieved the dry heat. Greeting me on the east-facing sidewalk were new wrought iron gates with curved lattice-work in a flower pattern that softened the metal. I walked through into a lush courtyard, and a wave of nostalgia overcame me. I reminisced over summer work parties and potlucks and smiled as I made my way through the village. It was quiet this afternoon; no doubt, villagers were out working, running errands, or enjoying the sunshine.

As I looked about I saw that some things had changed, while others remained the same. To the left were the familiar geodesic domes and to the right the support house, with the gray-water fishpond ahead of me. Past the pond and parallel to the woodshop, I saw a middle-aged bearded man in the process of building what turned out to be a new outdoor kitchen.

I had been walking for a few moments before Ralph emerged from the woodshop and spotted me. Ralph, still the main landowner, was a memorable man. He stood about a half a foot taller than me, now in his mid-sixties, with kind blue eyes, a gentle demeanor, and a lot of perseverance. As was common practice years ago, I had arrived unannounced, so Ralph asked about the nature of my visit. When I reminded him that I'd conducted research there a

Surviving Collapse. Christina Ergas, Oxford University Press. © Oxford University Press 2021.
DOI: 10.1093/oso/9780197544099.003.0007

dozen years ago, he said he remembered my face. But of course, many young people have come and gone in the years since. I explained I was just finishing my book and wanted to revisit the community that had inspired it in the first place.

With that, Ralph took me on a stroll around the property. Warning me to watch out for stinging nettle, he pointed out several new structures, including a 4,000 gallon rainwater catchment cistern that doubles as a star-gazing observation deck as well as a community meditation hut open to anyone in the surrounding neighborhood. I noticed that the food garden beyond the meditation hut, nestled closely to the naturally built triplex, was still flourishing with produce, such as summer squash, grapevines, and fennel.

After the short tour, Ralph invited me above the woodshop to talk. His home was as I'd remembered it when I'd interviewed him so many years ago. We entered through the kitchen and turned right into the living room and dining area. The wooden building had a homey, earthy smell and was lit by a skylight and south-facing bay windows, which overlooked the new outdoor kitchen.[1] We sat at a wooden table and launched into a wide-ranging discussion about the future of the earth.

We discussed climate change, the promise of a Green New Deal, and the grim possibility of civilization collapse. Ralph's sardonic sense of humor emerged as we deliberated over somber topics. He divulged that he was preparing for future food shortages and that was in part the motivation for the new outdoor kitchen under construction. He noted the dual fossil-fuel crises: climate change and peak oil. Of course, the burning of fossil fuels is the primary source of global CO_2 emissions, which contribute to climate change. He also argued that peak oil—the point when global petroleum supplies max out—has already occurred, and that eventually fossil-fuel shortages will cause global food shipments to come to a halt. The ecovillage can use the outdoor kitchen for canning and preserving operations to store enough food to feed the thirty ecovillagers and surrounding neighborhood for at least some time when impending collapse occurs. He knows that the food will not last long given that many of his neighbors aren't prepared at all, but his community-oriented mind, and perhaps fear of robbing mobs, has prompted a desire to build local community food security to prepare for a lack of global trade.

Whether or not peak oil has occurred is debated among experts, but many argue that peak oil is the reason why fossil-fuel companies are investing in less efficient extraction technologies, such as deepwater ocean drilling as well as the hydraulic fracturing of shale oil. Regardless, Ralph believes, as do many US politicians and economists, that the energy and material production transition needed to stave off climate change will devastate our economy and may well be an impossible undertaking to "keep this version of civilization going." He said that the single biggest source of his gloom is that "average people and our 'leaders' just don't understand how difficult it's going to be for wind and solar to replace the ease-of-extraction, the convenience-of-use, and the energy density of fossil fuels." Many products also rely on fossil fuels, not only for the energy in their production, but in their actual material construction, including plastics, asphalt, and lubricants as well as household items like lotions, clothes, and even medicines.[2] In response to these crises, Ralph repeated something he'd said in our first interview, years before: sustainability will come, with or without humans.

We talked for about an hour before Ralph said that he needed to return to work. He generously offered me use of the meditation space; he believes that more people need to meditate for a change in consciousness. After we said our goodbyes, I took him up on the offer.

Once I'd made my way to the meditation hut, I basked beneath its skylight. I was struck by the stark contrast between this verdant and lavishly overgrown place and the persistent droughts and famine in Guatemala forcing farmers to migrate that I'd just been reading about. I had trouble reconciling the existence of this thriving community in a world full of climate-related disasters. I know many people struggle with this dissonance. But what Asaṅga symbolizes to me is the possibility that ideal communities can be built, and must be, especially in the face of potential collapse.

I continue to reflect on my conversation with Ralph. Our discussion was disquieting, to say the least, although not unexpected. His bleak view of the earth's future is by no means unique. Our conversation came at the heels of numerous media headlines warning of civilization collapse.

C.1. Near-Term Civilization Collapse

The most recent coverage emerged after the release of a policy report by the Australian Breakthrough, National Centre for Climate Restoration, projecting that in a business-as-usual approach to dealing with climate change, which includes doing nothing to mitigate it and continuing to grow our economy, humanity has about thirty years until civilization collapse

(Spratt and Dunlop 2019). According to the report, German atmospheric physicist Hans Schellnhuber "warns that 'climate change is now reaching the end-game, where very soon humanity must choose between taking unprecedented action, or accepting that is has been left too late and bear the consequences' . . . if we continue down the present path 'there is a very big risk that we will just end our civilization. The human species will survive somehow but we will destroy almost everything we have built over the last two thousand years'" (cited in Spratt and Dunlop 2019, 6).

Other professionals have garnered recent attention for making similar or more ominous projections. University of Arizona professor emeritus of ecology and evolutionary biology Guy McPherson (2018) is known for his idea of "near-term human extinction," which he believes is likely by 2030. Sustainability leadership professor Jem Bendell's self-published academic paper, "Deep Adaptation," has gained recognition because it implores readers to "reassess their work and life in the face of an inevitable near-term social collapse due to climate change" (Bendell 2018, 2). David Wallace-Wells' 2019 book, *The Uninhabitable Earth: Life after Warming*, is an aggregate review of climate change–induced stressors that threaten hell on earth, including rising heat-related deaths, lower crop yields, increasing plagues, and warfare. Headlines have also chided that MIT researchers "predicted the end of civilization almost 50 years ago" (Dockrill 2018), because researchers from MIT created a computer program in the early 1970s called World1, the precursor to World3, which inspired the publication of the influential book, *The Limits to Growth*. The purpose of this program was to simulate the actual trajectory of human civilization versus other more sustainable trajectories. It then predicted—and recent evaluations still predict—that if current growth patterns continue, civilization is likely to collapse by 2040 (Turner 2014).

Tales of apocalypse and visions of post-apocalypse permeate pop culture as well. Not only do science-fiction writers imagine both utopian and dystopian futures, but podcasts challenge us to envision alternatives. In a podcast called *How to Survive the End of the World*, sci-fi writers Autumn Brown and adrienne maree brown explore the opportunities that could come with civilization collapse. They discuss how to imagine a post-apocalyptic world where everyone's needs are met with minimal power imbalances. They examine how the history of the last several hundred years was written by the victors—those who colonized, massacred, enslaved, and exploited whole peoples and landscapes. They envision utopian futures that integrate perspectives from peoples who have already experienced apocalypse at the hands of these "victors"—the African diaspora and Indigenous communities who have survived colonization, displacement, genocide, land dispossession, and slavery. These communities have proven resilient and offer important stories of survival and hope. They also reveal powerholders' tenuous positions. Even though the "victors" sought to kill off and control whole peoples and environments, they ultimately failed, because these communities prevail. And, they speculate, when civilization collapse occurs, the previously colonized may rise victorious.

Stories of survival and thriving communities may provide important insights into how societies might prepare for what may come. For my part, I don't see the utility in debating the timeline of civilization collapse. Although scientific models have become more sophisticated in recent years and are able to predict with greater accuracy what problems will arise, projections tend to be somewhere between too conservative or too extreme to be of practical help for planning purposes (Barnosky et al. 2014). I think the more productive conversation is to explore the way to live *now* that will allow us to transition into an uncertain future. Some of the more extreme collapse predictions exist under business-as-usual conditions; adapting communities now will serve both mitigation and resilience functions. Even as mainstream politicians and leaders resist discussing how to prepare for the disastrous effects of climate change, many experts agree that we need *unprecedented action* on a *global scale* to meet our climate challenges (Järvensivu et al. 2018). Some of these changes would have to be to be agreed upon by international organizations and governments. But many of them can and must begin on a local level, in our own communities. Reassuringly (or not), the history of human civilization collapse suggests that most societies do indeed survive and adapt, if in smaller numbers (McAnany and Yoffee 2010). Those most able to adapt are successful because they are prepared (Campbell 2018). The problem of social inequity is that the affluent have the most resources to put toward their communities' protection and resilience. For those of us who lack wealth, in order to prepare we need to invest in each other, our human and community resources, and especially in the poorest communities.

C.2. Beyond Collateral Damage and Sacrificial Zones

Because the root causes of ecological problems are social, sociological analyses offer valuable insights toward more sustainable solutions. Environmental social science research increasingly points to the need to address social inequalities in order to begin resolving environmental problems. Moreover, research strongly suggests that caring for those at the bottom of our social hierarchies actually serves to lift everyone else up in turn, similar to the fish rituals in the previous chapter. Unfortunately, some powerful world leaders continue to justify and minimize suffering in the name of human progress as "collateral damage" and "sacrificial zones." Among our casualties are the indigent, the unhoused, drug addicts, and victims of war, as well as endangered species of fauna and flora, decimated environments, and declining habitats. Perhaps at one time in humankind's development, in purely utilitarian calculus, the benefits outweighed the costs—we were finding cures for diseases, building cities, growing more food. But such calculations are no longer rational; in fact, they may be threatening human progress. As climate-related disasters grow, so do poverty, mental and physical health problems, homelessness, addiction, habitat loss, and species extinction. These losses threaten all the gains humans have made in the last few centuries as well as each of us individually (UN Human Rights Council 2019). In order to move forward, we must nurture, not torture, and restore, not destroy, those at the bottom of our social hierarchies.

To use the example of addiction from chapter 1 again, Portugal's experience provides useful insights. In the late twentieth century, Portugal faced an addiction crisis. One percent its population was addicted to heroin, and it had the highest rates of HIV infections from needle sharing in all Europe. The country had chosen to fight drug users, similarly to the US war on drugs, and nearly half its prison population was inside for drug-related offenses (Bajekal 2018). When state leaders recognized that they were losing the war, and drug addiction, overdoses, and disease infection rates from shared needles were increasing, they knew it was time for bold changes. Consulting with experts worldwide, Portuguese authorities came to understand that drug addiction is a symptom of past trauma and social isolation. In 2001, Portugal became the first country to decriminalize the consumption of all drugs. But they didn't stop there (Bajekal 2018). In fact, the most significant changes came from their drug treatment plans that focused on prevention, harm reduction, treatment, social reintegration, and rehabilitation.

What Portugal did that was indeed radical was a shift in perspective, from treating drug addicts like criminals to treating them with compassion, dignity, and care. The state sent, and continues to send, nurses, social workers, and psychologists to the most dangerous places with care packets—containing clean needles, condoms, and motivational messages—to offer treatment and services. Drug addicts went from feeling shame and societal stigma to knowing they had support services and a way out of addiction. As a result, Portugal's addiction and HIV infection rates have declined, and drug-induced deaths are among the lowest in Europe (European Monitoring Centre for Drugs and Drug Addiction 2015, 2018). The point is that the people of Portugal chose to care for those usually forgotten by society. Doing so has saved lives and money, as their public healthcare expenditures have dropped by almost 20 percent (European Monitoring Centre 2015).

Helping addicts is not the only example of how lifting up those at the bottom of the social hierarchy serves to lift everyone else up. The Housing First strategy, a model adopted in the past two decades by a number of public agencies to house chronically houseless people in places such as the state of Utah, significantly reduces public costs for crisis services, such as jail and emergency room visits, which cost taxpayers an estimated $30,000–50,000 per person annually (McEvers 2015; US Interagency Council on Homelessness 2017). Considering the environment specifically, inequality not only leads to disparities in environmental exposures that disproportionately burden the disadvantaged, but also to higher overall levels of exposure to health-damaging pollutants for everyone (Cushing et al. 2015). Indeed, many studies have concluded that the more unequal the society, the less healthy everyone in that society is in general (Subramanian and Kawachi 2004; Cushing et al. 2015).

Taken as a whole, recent research overwhelmingly demonstrates that caring for those at the bottom of our social hierarchies actually serves to lift everyone else up in turn. Singular people are *not* independent rugged individuals capable of mastering each other and the world on their own. Each of us is a small part of an interconnected web of relationships, and every part matters. We need the talents, insights, and perspectives of women, people of color, and Indigenous peoples just as much as that of white men. We need endangered insects who pollinate our food crops and supply sustenance to the small creatures that feed larger animals. We need thriving ecosystems to cleanse the air of pollutants, filter water, and provide nutritious soils. If as a global community, we choose to see, hear, and care for our most vulnerable human populations as well as our withering ecosystems, species threatened with extinction, and sacrificial lands, then we are likely to see positive changes. But we have to start somewhere, even somewhere small, like a neighborhood, town, or region. Both of the cases in this book demonstrate ways to build community capacity toward resilience. Their stories model how communities might more-than-survive collapse, because they are both thriving still. If building an ecovillage or urban farm is too tall an order, the following section provides smaller scale tools to help guide anyone looking to establish radical sustainability in their own community.

C.3. Radical Sustainability: Transforming and Regenerating Community

Radical sustainability is at once socially and ecologically transformative—dismantling hierarchies toward total liberation—and regenerative—healing and restoring the health of people and the planet. As the previous chapters illustrate, the capitalist project of sustainable development has prioritized the market above all else, which has proven incapable of addressing urgent socioecological crises. We've witnessed how the global aim of growing GDP has widened social inequities and devastated environments. This failing market supremacy is an important reminder that in socioecological systems, everything is interconnected. No one part can be prioritized at the expense of others for long. The task of challenging capitalism, and the many oppressive systems it reinforces, may be daunting, but I believe that having this information is ultimately empowering. By attending to and healing the social rifts that capitalism perpetuates, we can move toward a radical sustainability. Toward that end, we must reorient our attention to rebuilding neglected communities and ecosystems. We must not premise our social relationships on hierarchies or attempts at physical and psychological control. Rather, we must invite cooperative, reciprocal, and loving relationships into our communities, and we must extend this transformation to our nonhuman, or more-than-human, communities. A radical sustainability is informed by some of the following transformative and regenerative tools, which I describe in more depth below: a care narrative, community care, radical collective healing, community capacity assessments and building, skill sharing, and network building with other communities and organizations.

C.3.1. Social Transformation

We need social transformation, by which I mean large-scale social change that transforms systems of inequities into a nonhierarchical, equitable system. Part of this work is to transform our current narratives of abuse—based on control, fear, and legitimated violence—into narratives of care. Feminist activists and scholars have called the politics or ethic that should guide us the "care ethic" or, sometimes the "love ethic." bell hooks (2000, 87) asserts that "the underlying values of a culture and its ethics shape and inform the way we speak and act." She proposes we move toward a love ethic, because "awakening to love can happen only as we let go of our obsession with power and domination" (87). She further acknowledges that "embracing a love ethic means that we utilize the dimensions of love—'care, commitment, trust, responsibility, respect, and knowledge'—in our everyday lives" (94). Engaging in the ethic of care means that we attempt to leave things better than we found them. From the care ethic, we can begin to restore, revitalize, and regenerate local ecologies, and eventually, the biosphere. But this ethic must first imbue the stories and larger cultural narrative with care.

Social transformation toward the care narrative can occur in any social institution and may happen piecemeal, but it is more likely to be effective if it takes on a deeper cultural meaning. This transformation could include, but is not limited to, creating meaningful and creative spaces for those suffering in the least desirable and most dangerous work; getting rid of the most dangerous and undesirable work because most of it is likely not good for the environment either; participating in direct democracy in matters that concern one's life, work, and community, so that no one is ever surprised by changes underway; caring for the sick and elderly and integrating the perspectives they have to offer; treating everyone who falls ill to the best of the community's ability no matter their means; educating everyone, regardless of ability or resources, so that they are able to navigate their social and productive world; valuing local knowledge situated in both ecologies and communities; and maintaining egalitarian relationships at home, in the community, and at work, so that no one person owns all the resources that others depend on. Many of these changes likely would occur organically if we applied the care ethic to our communities, because it is necessarily relational.

Activists are challenging notions of individual self-care and are replacing it with community care. As we've seen, independent and individualist narratives do not serve us and may be in part responsible for humans' destructive relationship to the planet. Rather than believing we should own and control the world, turning our focus toward reciprocity and mutual care is what we need to heal. An example of how to move toward reciprocity is by advancing community care. Nakita Valerio (2019), a Canadian Muslim woman and community organizer, had a social media post go viral when she critiqued individualist notions of self-help for failing struggling communities battling institutionalized forms of oppression. Rather than bubble baths and pedicures as remedies for emotional distress, community care involves others stepping up to help out when individual community members are strained. She explains:

> Women of colour . . . have been calling for a shift to community care for years, and yet the onslaught of the self-care industry continues . . . community care is focused on the collective: taking care of people together, for everything from basic physical needs to psychological and even spiritual ones. Community care is a recognition of the undeniable cooperative and social nature of human beings and involves a commitment to reduce harm simply through being together. . . . Self-care only offers temporary relief to the deep-rooted structural challenges many of us face, and often it can be its own form of commodified labour that becomes an additional stressor for the person expected to perform it. So what does the alternative look like? . . . Community care means showing up; it means that when you find yourself in the position of being able to give more than you need to receive, you do so . . . this might mean receiving messages from someone who needs to be comforted and heard, bringing dinner to a sick friend or packing up an abused friend's belongings as part of their exit plan from domestic violence. . . . At different times and for different communities, people who are intentionally invested in caring for the communities they belong to might find themselves doing all of these things, or something else entirely. . . . Ultimately, community care is a commitment to contributing in a way that leverages one's relative privilege while balancing one's needs. It's trusting that your community will have you when you need support, and knowing you can be trusted to provide the same. . . . Unlike self-care, this response isn't about a bandage solution: It's about healing wounds together and eliminating the hazards that caused them together. . . .

Valerio notes that recognizing systemic forms of oppression, such as racism, as a threat to public health and advancing community care has healing potential. Again, the point here is to care for the most vulnerable and in need in our communities to help each other heal. Community care is an important feature of social transformation and regeneration. Part of community care work is allowing the space for healing from collective and individual traumas, especially as a result of oppressive narratives of abuse. Work toward transforming the abuse narrative to one of care will help to bring about nourishment, regeneration, and healing in communities and ecosystems, which will feed back into more transformation.

C.3.2. Regeneration

The *Merriam-Webster* dictionary defines regeneration as the process of "spiritual renewal or revival" or "renewal or restoration of a body, bodily part, or biological system (such as a forest) after injury or as a normal process" (*Merriam-Webster Dictionary* 2019). I use the term to differentiate it from restoration, which already has a specific ecological meaning. I also appreciate that it can be applied to the emotional, physical, or spiritual states of individuals, communities, and ecosystems. Ultimately, regeneration is about healing and restoring conditions that enable us all to thrive, especially communities and ecosystems most marginalized and exploited by the world system.

Counseling psychologists French et al. (2019) argue that systemic forms of oppression cause community-wide mental and physical harm and thus require collective approaches to healing. Radical healing is part of a process that centers on the strengths and resilience of oppressed communities. It first calls for developing a critical consciousness or awareness of systemic forms of oppression; then for allowing this consciousness to inform action against institutionalized violence; and, finally, for proactively working to prevent recurring trauma for their communities. Collective grieving and healing are also necessary for people fighting for climate justice. Some organizations, such as Good Grief and 350.org, are taking on collective healing efforts, especially in relation to climate change, by providing support groups and ten-step programs that help people move beyond debilitating grief toward action (Good Grief Network 2019). While collective healing is crucial within communities affected by institutionalized violence, action is needed beyond internal community work, as external factors will continue to affect them.

This external work can be conceived of as a form of regenerative community care, whereby allies and accomplices share in the responsibilities of recognizing systemic forms of oppression, educating others, and fighting against them (for more on allyship, see Kendall 2003; Indigenous Action Media 2014). External work also means that we recognize our privileges and begin listening to those without said privileges. Listening can happen by reading more about different forms of oppression; watching how people of color, women, and LGBTQ people (among others) are treated; listening when friends and acquaintances express grievances; and questioning how privilege influences our experiences of different spaces. While listening, it is important to recognize feelings of guilt and defensiveness for what they are: forms of self-protection wherein one perceives an accusation and thus reacts in such a way as to position themselves as the victim. This prevents the defensive party from hearing someone else with compassion. To move past these feelings, it's necessary to understand that others' experiences of systemic oppression are not about each of us as individuals, but rather are about larger societal structures that inhibit some peoples' ability to thrive while facilitating others'. Defensiveness is insidious and creates real barriers to communication (Brittle 2014). Working on our own defensiveness is crucial for recognizing that others' grievances aren't about us and allowing us to maintain care and compassion while giving others the benefit of the doubt. Total liberation also requires that we recognize that animals and plants are among the least privileged that need our attention as well.

If people are actually listening, they will hear and see the injuries of racism, heterosexism, classism, and speciesism (among others). Hearing and seeing are important first steps, but of course more action is necessary toward healing. We must acknowledge the ways we've been complicit in the systemic forms of injustice. An ethic of care requires that we all take responsibility for our part in these injuries, even if we intended no harm, and it means that we each tend to these wounds and nourish each other. There are many ways to tend to the wounds of exploitation. We can work to restore dwindling ocean mangroves and marshes that protect coastlines from storm surges. We can plant trees for reforestation. We can preserve endangered species' habitats. We can engage in radical love, hope, and community care by participating in mutual aid projects that share and (re)distribute resources to those in need. We can listen, validate, and offer compassion to those expressing their injuries of oppression. We can reject hierarchical relationships toward total liberation. We can stand up against institutionalized injustices toward animals, LGBTQ communities, people of color, and women by supporting movements, such as Black Lives Matter or Me Too. We can support national political and economic reforms that include a socially just transition toward a green economy, such as the Green New Deal in the United States. We can support climate justice movements striving for transformation, such as Indigenous Climate Action, Movement Generation, and Extinction Rebellion. Aside from fighting structural forms of oppression, another way forward is positive community development.

Positive community development is a part of the regeneration and transformation process that includes a capacity assessment aimed at building community strengths and bridges and healing community trauma. Capacity assessments are tools that help communities gauge what assets and needs the community has. This allows members to identify what will facilitate or potentially hinder movement toward community goals. Community members may choose to work toward improving weak areas so they are capable of remaining resilient in the face of unknown hazards. Community development is an iterative process, and requires communities to reassess their priorities and strengths regularly (UN Development Programme [UNDP] 2008; Stuart 2014). However, this action should not be done without a larger vision for systemic change, because large-scale social structures still inhibit community action. Capacity building may include local action toward designing regenerative food production systems that sustain the local community; direct democratic decision-making; attending to individual emotions and needs; healing from collective traumas; building local cooperative businesses; and restoring community ties so people can rely on each other, rather than commodified labor, for their needs. The important thing is that communities collectively make decisions about how to further develop their resources, skills, and talents.

Restoring capacity includes retaining and improving local knowledge and skills as well as building community networks for resources and skill sharing. Restoring capacity should build skills and confidence for individuals and groups; expand the leadership base; enhance community decision-making and problem solving; create a common vision for the future; help implement strategies for social change; and promote social justice and inclusive community participation (UNDP 2008; Stuart 2014). Reskilling and skill sharing are easy ways to implement community care in one's community and can take many forms. For instance, at the ecovillage, individuals teach permaculture workshops to people from both within and outside the village interested in regenerative agriculture. They also hold community events, such as potlucks or live music performances. Some communities have begun holding "fix-it clinics" where members bring broken household items and appliances and learn how to fix them. In the end, they have gained valuable skills and have functioning appliances! Other forms of skill sharing may include reading groups on communication techniques, such as nonviolent communication. Some communities have created ways to share other resources, such as lending libraries for tools and appliances so that not everyone needs to own a vacuum cleaner or lawnmower. In some neighborhoods, neighbors have mutually agreed to share space and resources by tearing down their fences and building community vegetable gardens in their own backyards open to all of their neighbors.

The people at el Organopónico in Cuba and Asaṅga in the Pacific Northwest approached aspects of the care narrative differently based on their unique cultures and political economies. The ecovillagers recognized that we are all traumatized by cultural expectations and individual experiences, and they engaged in individual and group forms of counseling to process that trauma as a community, in addition to sharing food, tools, and knowledge. Cubans, as a result of their revolution, are already ensured that everyone's basic needs are met, such as housing, education, healthcare, and childcare. At el Organopónico, farmers are cooperative members and thus all have a stake in the farm's productivity. As such, all members vote on what directions the farm will take. With their needs met, farmers can spend their energy on nourishing the soil and making the best organic produce to nourish their surrounding community—neighbors who nourish the farm by patronizing it.

Reassuringly, as the above examples illustrate, people have always envisioned and fought for alternative social relations and ways of being, and they continue to do so. Learning from examples like Asaṅga and el Organopónico can help us plan for alternative pathways toward sustainability. We have an opportunity to collectively reorganize in ways that are more socioecologically just, in order to combat the looming crises. With a solid vision, we can move beyond debilitating fear and denial toward a just transition to a new society with new goals, such as mutual nourishment rather than individual consumption. We can build a radical sustainability.

C.4. Building the Future

Twelve years ago, I began this project with a hope that is waning. As a US citizen, I expected then that the people of my country (one of the world's top polluters) would understand the implications of climate science, that our government would act based on the evidence, and that we would soon build communities able to mitigate, adapt to, and thrive in a changing climate. I hoped that even if my government failed to act, the people of the United States would rise up and demand action and that eco-communities, like Asanga, and urban farms, like el Organopónico, would proliferate in such numbers that they would overtake the average neighborhood. I believed that communities could and would pull together to find solutions to the urgent and escalating problems of a changing climate. While some communities do tirelessly fight for change, the events of this past decade have been deeply disillusioning. As we become further socially and politically fragmented, carbon emissions rise,[3] natural disasters worsen, temperatures increase, and climate-related fatalities grow. I find myself discouraged by the inaction of my federal government and the limited action taken by developed nations across the globe. And I'm not alone.

Climate scientists such as James Hansen have been warning leaders and the public about climate change-related risks for decades. A growing number of scientists and experts who study climate change are admitting to feeling distressed, hopeless, and defeated by the grimness of their research results and government inaction (Yong 2017; Richardson 2018). In the United States, people generally are experiencing increasing rates of climate-related anxiety, despair, and grief (Coyle and Susteren 2011 Nature Climate Change 2018; Obradovich et al. 2018). Psychiatrists warn of the exhaustion and burnout that I and many in my field of study currently suffer from (Clayton et al 2017).

The cynicism has set in for Ralph as well. Near the end of our already intense conversation, the talk took an unexpectedly dark turn. This man whose work I so admired, who devoted his entire adulthood and all his energies to building positive community, confided that he anticipated as society began to collapse and food scarcity set in, his ecovillage might come under siege. He can't help everybody, and he may be inundated by people who demand his skillset. Or worse yet, armed, roaming thieves may appear and steal all that he and the villagers have built. So peace-loving Ralph, avidly anti-gun his whole life, has bought himself a gun.

Nevertheless, I haven't lost all hope.

My greatest goal in writing this book is to motivate other people to pick up where I've left off. I now turn to those of you who want to act. I hope the lessons offered here will give you the inspiration and tools necessary to research, build, and explore other types of socioecological communities. I hope that communities will do the work of collectively envisioning utopian ideals and putting them into action. And I hope that social scientists will continue to assess these alternative socioecological ways of relating to see what is scalable. We can choose to stop destroying the world, cooperate with our fellow humans, and avoid war, or we can traverse farther into this burning hellscape. Whether it be through centralized state planning, mass grassroots mobilization, or civilization collapse, change is coming—and we are better off intentionally building a future together now.

APPENDIX

Methods and Cases

A.1. Methods

I conducted field research at two model sites, one I call Asańga, the US ecovillage, and the other I refer to as el Organopónico, the Cuban urban farm (all names are changed for confidentiality). At each site, I employed an ethnographic approach, defined as "first-hand participation in some initially unfamiliar social world and the production of written accounts of that world by drawing upon such participation" (Emerson, Fretz, and Shaw 1995, 1). The method I used was a qualitative cross-national (or cultural) comparison. This entails "studying certain phenomena in different cultural settings . . . beginning from the level of local practices, people's everyday life and experiences" (Gómez and Kuronen 2011, 685). I used this method in order to identify the real-world limitations to urban sustainability, especially in different cultures and political and economic contexts. This way, I could get a better picture of what is culturally specific and how to plan for the different problems that can occur under different political systems.

The specific phenomena I focused on were the attempts at developing urban sustainability at each site, the urban ecovillage and the Cuban urban farm. The first case I examined was an urban ecovillage in the Pacific Northwest United States. In the second case, I worked on an urban farm on the eastern periphery of Havana, Cuba. Many researchers have utilized the cross-national comparative case method to determine what is generalizable about the phenomena under study, what is culturally specific, and how the cases relate to one another (Gómez and Kuronen 2011). As an example, Rudel (2009) examines two international cases to show how people transform landscapes and the interrelated problems associated with deforestation in the Ecuadorian Amazon and suburbanization in New Jersey.

I applied a variety of techniques in each of my cases. Specifically, I conducted participant observation, semi-structured in-depth interviews, informal interviews, and observations. I analyzed media related to my cases from newspapers, television, art, and documentaries. I also read each site's official and unofficial documents. I kept detailed field notes of my observations, jotting them down at the site and typing them up later the same day. I recorded semi-structured interviews on a recording device, and I uploaded all interviews onto my password-protected computer and transcribed them for analysis. My interviews generally ranged from thirty minutes to three hours.

Dorothy Smith (1987) contends that in order to understand constraining institutions and power relations, it is important to ask respondents about their everyday actions to see how institutions organize these actions. Thus, in my interviews I asked respondents to describe their daily activities. I go into depth about each case below. First, I outline some advantages and disadvantages to qualitative cross-national comparative research.

A.2. Methodological Challenges

Qualitative case studies in general, and cross-national comparisons in particular, provide a richness of data that afford us a deeper understanding of a phenomenon that cannot be easily achieved with quantitative methods. The case method allows researchers to examine the particularities of certain phenomena in their unique cultural context. When comparing cases, researchers can glean what is specific to each context and what transcends cultural differences (Gómez and Kuronen 2011). Systematically analyzing case studies can help to develop theory by uncovering patterns, nuance, and latent features of a phenomenon. Immersive techniques give the researcher holistic insights by allowing them to experience what subjects experience in real time and space (Berg 2007). We learn through language what is culturally meaningful, prioritized, dismissed, or absent (Gómez and Kuronen 2011). In addition, we can parse out the discrepancies between normative prescriptions and everyday practices to reveal internal contradictions or legacies of colonial intrusion (Burawoy 1998; Gómez and Kuronen 2011).

As a researcher foreign to Cuba, some challenges I faced included understanding the idiosyncrasies of a new culture and participants' conceptual frameworks, which affect how they both interpret and answer interview questions (Øyen 1990; Gómez and Kuronen 2011). I attempted to get around these problems by using a comparative case study from the United States to examine the different forms of language that respondents use, word choices, and priorities in each context, as well as the meanings of words, concepts, and phrases.

A.3. Cases

A.3.1. Significance of Each Case

If we are to critique the global project called (however ironically in hindsight) sustainable development, and argue that another way is necessary, we should be looking to emancipatory projects already underway to highlight alternatives (Wright 2010). And, if we are to undertake emancipatory projects, we must be aware of the types of problems people encounter so that we can plan accordingly. Thus, in my work I assessed the barriers to urban sustainability projects in order to uncover the ongoing challenges their practitioners face. I chose exemplary cases, or model sites, because it is through the most innovative projects that we can begin to see how far we still need to go to achieve sustainability. Model sites can impart important lessons not only by providing examples and inspiring ingenuity. They can also provide glimpses of systemic deficiencies or day-to-day activities that undermine their own progress. My goals are to develop a theoretical understanding of urban sustainability at a local scale through comparative case studies representing two different political and economic contexts. Model sites are the only examples we have of best practices, and they still face barriers, both institutional and interactive.

A.3.2. Ecovillage: Pacific Northwest United States

The first case study I investigated is an urban ecovillage, Asańga, located in a city in the Pacific Northwest United States. Ecovillages are a relatively new and burgeoning phenomenon internationally. The particular ecovillage in which I conducted interviews and participant observation is ideal to study because it was established during the conception period of ecovillages, the early 1990s. It still retains some original residents who experienced the emergence of this movement (Gilman 1991; Smith 2002b). A stated goal of the individuals at this site is to spread their vision of sustainability through community outreach projects, like educational tours, television shows, and local newspapers (Ergas 2010). The ecovillage is strategically located within an urban area where residents can attempt to change local city housing regulations, raise environmental awareness, and serve as a model of sustainable living for the local city. Sites like these challenge city versus nature dichotomies that inform traditional urban development policies (Čapek 2010). Villagers often find themselves constrained by the larger community because of slow-moving bureaucracies and some neighbors' resistance to change. Regardless of these constraints, the village has persevered. There are limitations to this case study; for example, it is not generalizable to all ecovillages or social movement groups. However, my data reveal everyday challenges that people in the United States face when trying to live sustainably.

I first encountered the ecovillage in the summer of 2004 and regularly visited the space over the course of several months. I went back years later to conduct research and, because I had previously spent the summer and fall of 2004 visiting the community regularly, I was able to reenter easily. Even after three years, the property owners and original conceivers, as well as its residents, still recognized my face and they met with other community members to reassure them that my presence would not be intrusive.

I spent slightly more than two months, from July 2007 to September 2007, visiting and living in the ecovillage community; interviewing, observing, and participating in community activities; and engaging villagers in discussion. In exchange for being allowed to sleep on a futon mattress in a teenager's living room, I helped out in work-trade, which included moving compost, cleaning rabbit cages, and doing domestic chores, and I became involved in some individuals' environmental awareness projects. After my stay, I continued to visit the community about once a month for the next six months. I promised interview subjects confidentiality. To honor this, I use pseudonyms for people and places throughout my analysis.

My twenty-four interviewees included twenty-three of the twenty-seven adults older than eighteen who lived at the ecovillage when I entered in early July. I also interviewed a woman who moved off the property a year earlier but had lived there previously for a total of three years. The ecovillage population is constantly changing, but the community is consistently multigenerational. My interviewees' ages ranged from nineteen to seventy-seven years with a mean age of thirty-six. Fifteen interviewees identified as female, and nine identified as male. Every interviewee was white, mostly Western European ethnics, a few Eastern Europeans, and a few individuals who claimed to have small parts of Native American ancestry. Of the twenty-four people I spoke with, twelve had lived there for at least a year or more.

For triangulation purposes, I conducted semi-structured, in-depth interviews and participant observation, and I analyzed written community materials. Interviews were recorded and later transcribed, and field observations were jotted down in a field notebook and typed up at the end of each day. Interviews lasted anywhere from forty-five minutes to three hours, and questions, which I explain in more depth below, focused on personal values, everyday actions including work and play, and reasons for living in an ecovillage (Ergas 2010; Ergas and Clement 2016). These questions, informed by the movement culture literature, gave me insight into personal and movement goals that ecovillagers strive to accomplish and how they work toward achieving these goals in day-to-day life (Melucci 1995; Smith 1987; Burawoy 1998).

I investigated interview data, field notes, and some written materials to decipher respondents' understanding of meaning and action and how meaning and actions are constructed within a dominant society that arranges their opportunities and constraints. I also asked questions about how individuals viewed the city and dominant culture to understand their critique better. For example, to interpret ecovillagers' perceived opportunities and constraints in their city, I asked, "What does a typical day look like? Is the larger community conducive to maintaining the ecovillage? And, what resources do you utilize from the city?" Often, personal goals are interconnected with movement goals, such as choosing the ecological brand of soap to use or specifically buying local produce. Respondents' answers allowed me to see what they perceived as favorable conditions, such as access to bike lanes, or structural impediments to their goals, such as having to pay for land and therefore work in the formal economy. Their answers also elucidated instances of agency where they found ways to go around or confront impediments to their goals, such as in an instance with one resident, Ears, who went door to door in his neighborhood in an effort to build more support and community.

I analyzed interview data, a welcome pamphlet, and villagers' everyday actions to distinguish their goals. The co-owner, Emily, wrote and edited a four-page, typed welcome pamphlet. Villagers agreed to new rules during meetings, and newcomers were given the pamphlet during their initial entry and interview process. When coding interviews, I first determined recurring themes from interviewees' responses to questions regarding ecovillage community values, personal values, their understanding of dominant cultural values, and problems they see both in the ecovillage and in dominant culture. Then I evaluated how this understanding of values translated into everyday action by examining responses individuals gave to questions regarding what a typical day looks like, what a typical day might look like if they did not live at the ecovillage, and how individuals' felt the ecovillage was affected by its location in a city. Finally, I used the welcome materials and my observations in my field journal to confirm, disconfirm, and contextualize interviewees' responses to my questions. Through these responses and notes, I disentangled the ecovillagers' most prominent goals and how they understood these goals translating into everyday action within the confines of the dominant culture (Ergas 2010; Ergas and Clement 2016).

A.3.3. Entrée and Power

A discussion of power dynamics is an important part of my reflection on how I gained entrée during my fieldwork. I was very aware of the power dynamics in my interactions with interviewees, more so in Cuba than at the ecovillage. Burawoy (1998) argues that effects of power are the biggest limitation to qualitative case-study research. Power relations are ubiquitous, but this does not mean that we should abandon the qualitative case-study method because of our concerns about the effects power will have on our research.

In particular, researchers cannot avoid dominating or being dominated by our participants. Indeed, Burawoy (1998, 22) contends, "Entry is a prolonged and surreptitious power struggle between the intrusive outsider and the resisting insider." Hierarchies based on gender, race, class, nationality, and position in relation to authority affect ideologies, interactions, and access to resources. Power effects can come up in respondents purposely silencing themselves by leaving out contentious information.

Finally, Burawoy cautions against normalization, or fitting cases and participants to theory and vice versa, which molds participants into our academic frameworks. He suggests that to temper this problem requires that we more closely embed our analysis in "perspectives from below, taking their categories more seriously, and . . . working more closely with those whose interests the study purported to serve" (1998, 24). However, researchers can learn from their interactions with participants; through these conversations, a reflective interviewer can uncover how local processes relate to social forces, like power dynamics based on systems of oppression and differences in understanding or meaning based on different statuses (Burawoy 1998; Gómez and Kuronen 2011). In order to deal with power effects, I give a detailed account of these dynamics, including a description of my positionality.

I want to acknowledge that my understanding of what occurred at the ecovillage is influenced by my situated knowledge, or assumptions based on my race, class, gender, and nationality, among other things (Harding 1991; Collins 2000). Feminist and postcolonial scholars have discussed what some consider the limitations of ethnographic research through power effects, or our embeddedness in social relations, between researcher and participants. Some scholars suggest that addressing these power differentials in a reflexive science, or acknowledging our situated knowledge and embracing our positionality, is actually beneficial to the scientific project (Harding 1991; Burawoy 1998; Collins 2000). Thus, I engage in a discussion of my particular standpoint, or my different situation based on the historical processes of colonialism, classism, racism, and sexism.

During my stay at the ecovillage, my positionality more closely resembled ecovillagers than my Cuban respondents. Similar to many of my ecovillage subjects, I am an educated white, nonbinary woman (who identifies as agender). I was working class (at the time and previously). Dissimilarly, I am a second-generation Cuban and Greek American, born to immigrant parents. I was ten years younger than the mean age of my ecovillage interviewees, though I interviewed many women close to my age. At the time, I had just begun my graduate program and was conducting research for my master's degree field paper. In these aspects of my identity, my respondents and I shared level positions in social hierarchies.

While I promised my respondents that I would share their stories, their words were filtered through my analysis. To deal with this power differential, I emailed each interviewee a copy of their interview transcript along with my notes. During each interview, I informed respondents that if they felt misrepresented or wanted to clarify something when they looked over the transcript, they could contact me with different information. I also emailed a copy of my manuscript over the community listserv before I attempted to publish it, to gauge the community's response. I received no response. I have published articles based on this research and honored many respondents' desire to have their stories told.

When discussing identity, it is important to examine social factors such as race, gender, class, and sexuality. In my observations, I noticed some interestingly gendered aspects of the community. In general, men thought women held more power in the community, and I observed some gendered work. Gendered work seemed exemplary of the embeddedness of ecovillagers in the dominant culture. I also want to acknowledge that I only caught a snapshot of this village. I am not aware of the typical makeup of the ecovillage. It is a transient space and may have looked very different a few months before I visited. For example, all my respondents were white. I do not know if this is typical of ecovillages in general, indicative of the homogeneity of the city, or a matter of the time I was there. Thus, I do not analyze race and ethnicity in this discussion. Additionally, sexuality was rarely brought up in my interviews. With regard to class, some of my interviewees came from poor or working-class backgrounds, although most were middle class. The fact that most identified themselves as middle class is consistent with Inglehart's (1977) description of postmaterialists who value quality of life over material signs of wealth, which may have been a sign of class privilege.

A.3.4. Urban Farm: Havana, Cuba

The second case study is of an urban farm, el Organopónico, located in a peri-urban area on the eastern side of the city of Havana, Cuba (Ergas 2013b, 2014). I first visited the farm in the summer of 2010 and went back to conduct more research from December 2010 to February of 2011. During my second trip, I conducted participant observation and semi-structured interviews with workers at the model farm. Studying Cuban urban agriculture is important because empirically investigating efforts toward sustainability occurring in the real world can help us understand the processes that facilitate and inhibit environmental reform. We can then examine the costs and benefits of achievements, the ongoing challenges, and the contradictions. In this way we discover possibilities that we are not able to imagine in our own context. It is important for scholars in the Global North, who are often sheltered from significant struggles abroad, to witness and report the efforts toward sustainability that are successfully being applied elsewhere.

Havana, Cuba is an ideal site to conduct research on urban organic agricultural practices, and it is theoretically interesting because Cuba is "recognized as a world leader in alternative organic agriculture and [urban agriculture]" (Premat 2005, 153). In Havana, locals grow 60–90 percent of the produce consumed in the area (Premat 2005; Stricker 2007). Additionally, the World Wildlife Fund deemed Cuba the only sustainable nation in the world in the 2006 Sustainability Index Report (Hails, Loh, and Goldfinger 2006). Cuba is also interesting because the Cuban government institutionally supports and subsidizes these agricultural practices, unlike the situation in the United States. The specific farm I observed is the model site featured in documentaries on Cuban urban agriculture and is where tourists are taken to observe urban agricultural practices in Havana.

Cubans can make a living doing urban agricultural work as this work is subsidized and supported by the Cuban government. However, even in ideal circumstances some problems persist. To try to understand limitations to sustainability, I made two trips to Havana, once in June of 2010 and again from December to February of 2011. My field research involved interviewing, working with, and observing scientists and farm workers at one site. I read formal and informal organization documents and saw documentaries put together by Cuban feminist organizations on Cuban women's work in agriculture. I also stayed with a farm-worker woman and her family.

During my first trip in the summer of 2010, I established contacts through Global Exchange, based in the United States, and Cuban organizations such as the National Urban Agriculture Group, the Federation of Cuban Women, and the Cuban Association of Agriculture and Forestry Technicians. In the winter of 2011, I went to Cuba for a final round of data collection, during which I worked on one urban farm. While there, I conducted many informal interviews with the men I worked with, and fifteen semi-structured interviews with thirteen women and two men. I recorded and transcribed the formal interviews and took detailed notes of my work observations and informal discussions with coworkers (Ergas 2013b, 2014).

A.3.5. Entrée and Power

My understanding of things that happened on the farm also was influenced by my situated knowledge, or assumptions based on my race, class, gender, and nationality, among other things (Harding 1991; Collins 2000). I more often confronted power dynamics in my day-to-day interactions with Cuban farm workers than I faced in interactions at the ecovillage, and different aspects of my standpoint are more salient with my Cuban respondents than the ecovillagers. These power differentials are based on the historical processes particular to Spanish colonialism, Western imperialism, and Cuba's unique cultural expressions of racism and sexism. To address these concerns, I reveal here that I am a white, Cuban- and Greek-American, US-citizen, nonbinary woman. My family immigrated to the United States from Cuba after the revolution and were part of an elite, white, land-owning class and tend to be conservative. I, however, am a sociologist from a university in the United States with left-leaning politics. I can speak Spanish and understand it, but my accent reveals that I am not native. These aspects of my identity undoubtedly affect my interpretation of events in Cuba, how Cubans perceive me, and what respondents choose to disclose with me. With my standpoint acknowledged, I will describe my interactions with the Cubans at the farm.

While the farm president was enthusiastic about my participation at the farm, as was his daughter, who housed me, some other workers were skeptical. However, most Cubans I encountered were friendly and eager to talk to me. This may have had to do with the fact that the president seemed happy to have me there and asked that people work with me. Some Cubans seemed particularly interested in talking to a foreigner, especially a US citizen who was interested in their work. The young people I encountered were enamored with American culture. Some disclosed that they longed to live in the United States. (It is not uncommon for young men and women, in particular, to marry foreigners in order to move away with them.) When I had an opportunity to divulge my Cuban ancestry to my coworkers, many of them would get excited and claim me as Cuban. However, some people seemed to have no interest in talking to me, either because I was a foreigner or otherwise.

I am aware of some of the privilege associated with being a Cuban-American with fair skin in Cuba. I have the right to freely enter and exit the country, for which Cuban citizens must obtain special permission. My socioeconomic class was higher than many Cubans, because I have access to US dollars, a more valuable currency than the Cuban peso (though some Cubans obtain remittances from expatriate family members that increase their class standing). Being white is a privileged identity in Cuba, as Black or mixed Cubans experience racism. I am from the United States, an imperial power that limits Cubans' access to resources, many of which I can easily access. Notwithstanding, there were times when my privileged identities were limiting within Cuba.

My experience dealing with government officials in Cuba was very different than my interactions with Cuban people on the farm and streets. Perhaps as a result of government censorship (Reporters Without Borders 2012) and/or terrorist attacks against Cuba, like Operation Mongoose initiated by the Kennedy Administration (U.S. Department of State Office of the Historian 1997), and longstanding political struggles with the United States, Cuban authorities are suspicious of foreign researchers, especially from the United States. Unless sponsored by an official Cuban organization, like a university, the Cuban government will not allow foreign researchers to investigate Cuban activities. Gaining sponsorship is a challenging, time-consuming, bureaucratic process. As part of the challenge, I was questioned at the airport while officials read through my research documents, proposals, and questionnaires.

In juxtaposition to my encounters with government bureaucracy, the atmosphere on the farm was very casual. The president often in jest told other workers that I was his niece. When I tried to refer to him in Spanish using the formal "you," he insisted that he was not above anyone nor was he that old, and thus I should use the informal "you." People on the farm referred to each other in informal ways and often called each other comrade, a gender-neutral and nonhierarchical form of address, likely a remnant from their ties to the Soviet Union.

Visibly being a woman (even if I do not identify as such) had its own set of problems. People on the farm often bantered in a flirtatious manner. Men and women often referred to each other as beautiful or handsome. On many occasions, men told me how beautiful I was, and I received invitations for love, marriage, and one-night stands. For a feminist enculturated in professional settings in the United States, I was uncomfortable with these propositions. Generally (although there are plenty examples of workplace sexual harassment), US professionals avoid discussing sexuality to prevent accusations of sexual harassment. However, in Cuba, my attempts to avert these topics proved impossible. In addition to unwanted propositions, men would refuse to allow me to do physically demanding labor even when I insisted on it. They noted my feminine stature and told me to let the men take care of the heavy lifting. Despite these dynamics, working on the farm was generally a delightful experience. The temperature was usually between 70 to 80 degrees Fahrenheit; it was almost always sunny; and most of my work could be completed outside in a cool breeze under the shade (Ergas 2013b, 2014).

A.4. Coding and Analysis

I applied both inductive analysis, emerging from my data, and deductive analysis, following from theory, to my field notes and interview transcriptions. I began by open, line-by-line, coding of my notes and transcriptions. I noted themes, patterns, and topics of interest. I wrote initial memos noting locations of ideas, themes, situations, and theories that I gleaned. As themes emerged, I went back and began more focused coding, looking for specific topics, themes, and variations. At this point, my memos began to deal more with comparisons, nuances that I had previously missed, and the evolution of my cultural understanding over the duration of my data collection at each site. I engaged sociological theories with the emergent ideas from my data (Emerson, Fretz, and Shaw 1995; Berg 2007). Through this process, I put together the specific analyses of each of my cases.

To compare my distinctive cases, I asked what the similarities and differences were in the language that Cubans and ecovillagers used; the ideology, assumptions, or conceptual framework they departed from; their actual daily actions and activities; and how they articulated what they believed they were trying to accomplish (Emerson, Fretz, and Shaw 1995). I looked for differences in culture, language, conceptualization, and semantic similarities that obscured cultural distinctions. For example, the word sustainability meant something very different to my Cuban respondents than it did to my ecovillagers. Ecovillagers' conceptual framework was more akin to that of environmentalists and social movement participants. However, my Cuban respondents saw themselves more as farm workers providing for their community. With these differences in mind, I wrote chapter 5, comparing the urban sociological experiments.

A.5. Semi-structured Interview Questions

A.5.1. Ecovillage

Demographic Questions

1. What is your birth date?
2. What gender do you self-identify with?
3. What is your ethnicity?
4. What is your occupation?
5. What level of schooling did you complete?
6. What level of schooling did your parents complete?
7. Do your parents/siblings own property?
8. What occupations do your parents/siblings inhabit?
9. Do you own property?

Life Questions

1. What events brought you where you are now?
2. What does your typical day look like?
3. What would your typical day look like if you were not living here?
4. How do you get around?
5. Where and/or how do you obtain food?
6. What does living in intentional community mean to you?
7. What is most important to you about living in intentional community?
8. What do you think are the most important issues in your intentional community?
9. What do you think are the most important issues in the larger community?
10. How are decisions made in the intentional community?
11. How does the community deal with a lack of resources, or resources running low?
12. How does the community deal with people who are not contributing?
13. How does the wider community perceive your community?
14. Is the larger community conducive to maintaining this community?
15. How do you sustain your living conditions?
16. How do you obtain money (or do you even need money)?
17. What resources do you use from the city?

Attitude Questions

1. What do you think are some mainstream cultural values?
2. How do you feel about these values?
3. Do you have any judgments, positive or negative, toward these values?
4. What are things that your community values?
5. How do you feel about these values?
6. Do you have any judgments, positive or negative, toward these values?

A.5.2. Cuban Urban Farm

Demographic Questions

1. What is your birth year?
2. What gender do you identify with?
3. What race/ethnicity do you identify with?
4. What is your occupation?
5. What level of schooling did you complete?
6. What level of schooling did your parents complete?
7. What occupations do your family members do?

Work Questions

1. How did you get involved with the work that you do?
2. What work do you do on a daily basis? Or, what does a typical day look like?
3. What work do you do before and after your day job?
4. What kind of work would you do if you did not do this work?
5. How many hours of work do you do each day? On each activity?
6. What is the most challenging work you do? How often do you do it?
7. What is the most creative or interesting work you do? How often do you do it?
8. What is the most mundane/boring work you do? How often do you do it?
9. Do you do the same kinds of work as the women/men in a similar position? What kinds work do you do that is different/same? What work do the women/men do?
10. What do you enjoy most in your work?
11. What do you enjoy least in your work?
12. Overall, is your work experience enjoyable? If not, why/how? If so, why/how?
13. What are the biggest obstacles you face to doing your work?
14. Do you experience men (or women) not taking your work seriously?
15. What facilitates your work experience?
16. What are daily problems you face, and how do you go about solving them?
17. How do you feel about sustainability? Organic agriculture? Urban agriculture?
18. How did the farm come about?
19. What is the history of this farm/organization?
20. What are the objectives of the farm/organization?
21. How do you fertilize the soil? Who does it?
22. How do you choose a farm site, and who does it?
23. What do you grow and why? Who chooses?

Notes

Introduction

1. The mayor of a small communist village. Dan Hancox, "Spain's Communist Model Village," *The Guardian*, October 19, 2013, https://www.theguardian.com/world/2013/oct/20/marinaleda-spanish-communist-village-utopia.
2. Much of our plastic waste will end up in the ocean and other waterways threatening wildlife (Hoornweg, Bhada-Tata, and Kennedy 2013).
3. Seventy percent of global oxygen is produced by oceanic phytoplankton; if it dies off due to ocean acidification, breathing at sea level will be like breathing at the top of the Himalayas (Sekerci and Petrovskii 2015).
4. Litfin (2014, 18–19) describes ecovillages as living laboratories and uses pioneer plant species as an analogy. Pioneer plants arise after an ecosystem has been destroyed by an event that leaves the ground difficult to inhabit. Pioneer species usually have strong roots, need little water, and generally grow fine in harsh environments. When they die and decompose, they make the ground fertile and more suitable for less hardy plants.

Chapter 1

1. Modernization consists of multiple concurrent processes that feedback into each other but began perhaps earlier than the sixteenth century in England with the enclosure of common land into privately held large farm landholdings; the enclosures caused the displacement of subsistence agriculturalists and pastoralists, which created an "unskilled" labor force dependent on industrial employment; displaced laborers propelled industrialization and urbanization as they moved into cities seeking employment, and they became urban consumers dependent on modern products for sustenance. The dual, yet interdependent, development of large-scale industrial agriculture and urbanization are the physical manifestations of technological advancement and individual freedom, in that heavy machinery relieves most people of farming duties and allows (or forces) them to pursue other work in urban areas. The displacement of rural communities is at once a cause and consequence of urbanization. To sustain urban environments, supporting rural developments are needed in the way of industrial agriculture, resource mining, and river damming. While urban land occupies between less than 1 to 3 percent of global landcover, the land area required to sustain urban populations is significant (Ergas, Clement, and McGee 2016). According to the World Wildlife Fund's (2014) Living Planet Report, we use 18.1 billion gha of bioproductive land, which is

about 50 percent more than the earth's available biocapacity, or the land actually available to provide ecological goods and services. We have been exceeding biocapacity for over forty years, and we are seeing the long-term consequences of overconsuming resources, including "diminished resource stocks and waste accumulating faster than it can be absorbed or recycled" (32). Though it is difficult to estimate urban areas' total resource requirements, the most developed countries tend to be the most urban, and they consume and dispose of waste at levels that are about five times higher on average than low-income countries.

2. Critics also show that capitalism ignores natural laws such as entropy. For an extended discussion of entropy and capitalism, see Rees and Wackernagel (1996).
3. Further, even though proponents of technological development assume that improving resource efficiency during production naturally leads to resource conservation, a body of empirical research has shown that a phenomenon known as the Jevons paradox, or the paradox of dematerialization (Jevons 2001 [1865]; Polimeni, Mayumi, Giampietro, and Alcott 2008), is a common phenomenon at the national level. Jevons was the first to describe the association between the declining ecological intensity (i.e., rising efficiency) of the economy and the rising total resource or energy consumption (York, Rosa, and Dietz 2003, 2004, 2009; York 2006, 2010; York et al. 2011).
4. High human development indicators are UN measures of life expectancy, educational attainment, and gross national income (GNI) per capita.
5. Some proponents of sustainable urban development believe that we can achieve urban sustainability through nongrowth steady-state economies (SSE; Commoner 1992; Daly 1996). Daly defines a steady-state economy by maintaining that "the aggregate throughput is constant, though its allocation among competing uses is free to vary in response to the market. Since there is of course no production and consumption of matter/energy itself in a physical sense, the throughput is really a process in which low-entropy raw materials are transformed into commodities and then, eventually, into high-entropy wastes. Throughput begins with depletion and ends with pollution. . . . Qualitative improvement in the use made of a given scale of throughput, resulting either from improved technical knowledge or from a deeper understanding of purpose, is called "development." An SSE therefore can develop, but cannot grow, just as the planet earth, of which it is a subsystem, can develop without growing. . . . The other crucial feature in the definition of SSE is that the constant level of throughput must be ecologically sustainable for a long future for a population living at a standard of per capita resources use that is sufficient for a good life" (31–32). This falls in line with the idea of "strong sustainability," as a non-growth-oriented economy should only be concerned with maintaining the livelihood of the population. Again, we have no examples of cities with high standards of living and steady-state economies.
6. In this section, I refer to social capital in the Putnam sense of the term, which stands in contrast to Bourdieu's use of the term. For Putnam, social capital—building networks of cooperation and reciprocity—is integral to a healthy civil society, whereas Bourdieu sees social capital as a means of maintaining inequality because elites maintain distinct and impenetrable social networks.
7. In *The Ecological Rift*, Foster, Clark, and York (2010) assert:

> This ecological rift is, at bottom, the product of a social rift: the domination of human being by human being. The driving force is a society based on class,

> inequality and acquisition without end. At the global level it is represented by . . . the imperialist division between center and periphery, North and South, rich and poor countries. This larger world of unequal exchange is as much a part of capitalism as the search for profits and accumulation. (47)

Though their focus is on class and divisions between developed and developing nations, these injustices are also perpetuated against most historically marginalized groups within nations based on racial and gender hierarchies as well as divisions in nationality, age, religion, and sexuality (Capek 1993; Taylor 2000).

8. The documentary was written by University of British Columbia law professor Joel Bakan. It is informed by and includes information from the DSM-IV, the mental disorders diagnostic manual. The psychopath checklist includes "callous unconcern for the feelings of others; incapacity to maintain enduring relationships; reckless disregard for the safety of others; deceitfulness, repeatedly lying and conning others for profit; incapacity to experience guilt; and failure to conform to social norms with respect to lawful behaviors."
9. Homo sapiens are thought to have emerged roughly 200,000 years ago, though this timeline is debated and varies from 40,000 to 315,000 (Rafferty 2020).
10. Innovations in production efficiency such as Fordism, characterized by Ford Motor Company's specialized assembly-line production in the early twentieth century, virtually made workers cogs in a machine.
11. As Fromm (1956) writes of our economic self: "He has been transformed into a commodity, experiences his life forces as an investment which must bring him the maximum profit obtainable under existing market conditions. Human relations are essentially those of alienated automatons, each basing his security on staying close to the herd, and not being different in thought, feeling or action. While everybody tries to be as close as possible to the rest, everybody remains utterly alone, pervaded by the deep sense of insecurity, anxiety and guilt which always results when human separateness cannot be overcome. . . . Automatons cannot love; they exchange their "personality packages" and hope for a fair bargain" (79–81).
12. Consumption of luxury items has dramatically increased during the last half century as well, as modern consumers consume twice as much as they did fifty years ago (Leonard 2010).
13. In 2009, environmental damages alone from the world's largest 3,000 companies cost about $7.5 trillion, about 13% of global economic output. Many industries would not be profitable at all if their environmental costs were fully calculated (Trucost 2013).

Chapter 2

1. I have changed the village name, like all names in the chapter, for confidentiality purposes. But Asaṅga, a reference to a Buddhist teacher, captures the spirit of the village's real name.
2. Some residents engaged in "dumpster diving," named for taking items out of dumpsters, but which consists not just of getting food from dumpsters but also of going to various grocery stores or bakeries. Many store employees are aware of dumpster divers, and

some will put all their old bread into large plastic bags to make it easier for divers. Other shop owners instead choose to lock dumpsters to keep divers out.

3. Rural ecovillages that disengage with larger communities to live their own vision of community are considered prefigurative movements.

Chapter 3

1. My main focus is on the social aspects of sustainability as they relate to socioecological practices, such as regenerative farming. I understand that this is an anthropocentric way to look at ecological sustainability, but it is my aim to highlight social practices that tread lightly on natural environments while maintaining high levels of human development.
2. Again, transformation means dismantling hierarchies toward total liberation and regeneration means healing and restoring the health of people and nature.
3. Their prosperous sugar industry allowed Cuba to build up its infrastructure. As a result, Cuba has been a highly urbanized country for most of the twentieth century. As of 2010, 75 percent of the Cuban population lives in urban areas (ONEI 2011). Havana is the most populous city in Cuba, with a population of 2,135,498, about 19 percent of the national population (ONEI 2011).
4. As of this writing, the embargo is still in effect.
5. Although Cuba's GDP has nearly tripled since 2000 (World Bank 2017).
6. All names have been changed for confidentiality purposes.
7. Rule of law includes property rights and freedom from corruption; regulatory efficiency includes business, labor, and monetary freedom; limited government includes government spending and fiscal freedom; and open markets include trade, investment, and financial freedom (Economic Freedom Index 2013).
8. CEDAW "defines what constitutes discrimination against women and sets up an agenda for national action to end such discrimination" (UN Women 2009).
9. In 2008, Cuba ranked 47 among nations according to the GII, but their ranking went down to 58 in 2011 and 63 in 2012 (UNDP 2013).

Chapter 4

1. This mantra is reminiscent of the Thomas theorem, advanced by sociologists in the 1920s, which suggests that ideas are real in their consequences.
2. A paradigm is a philosophical or theoretical framework within which theories, laws, and generalizations, and the experiments performed in support of them, are formulated (Kuhn 2012). Hegemony is a form of social control from one state over another, or in this case, one global region over others. This power is usually maintained through cultural, economic, political, and militarized means.

3. Val Plumwood's (2002) Master Model theory examines how this change in reason informed and anchored Cartesian hierarchical dualisms that privilege certain groups while simultaneously subordinating and oppressing others, including women, Indigenous peoples, people of color, sexually diverse beings, animals, and the environment. This Master Model operates in five interconnected ways through: (1) backgrounding: "the Master utilizes the services of the other and yet denies his dependency;" (2) radical exclusion: the "Master magnifies difference between self and other, and minimizes" similarities; (3) incorporation: "the Master's traits are the standard against [which] the other is measured;" (4) instrumentalism: "the other is constructed as having no purpose other than to serve the Master;" and (5) homogenization: "the dominated class of others is seen as uniform and undifferentiated" (Gaard 2017, xxiv). As Western, elite, white men ascended as the morally superior possessors of dispassionate reason, they thus entrenched the dualisms of mind/matter, active/passive, man/woman, white/Black, and civilization/nature that perpetuate sexism, racism, and anthropocentrism, or the centering of humans and their experiences above all others, as well as permeate our political and economic rational today (Merchant 1980; Plumwood 2002; Gaard 2017).
4. Volumes exist on alternative economic systems that are more sustainable than capitalism. Evaluating the economic aspects of sustainability is beyond the scope of this book because the purpose is to highlight the dynamic between the social and ecological. However, for more information about alternative economic systems, see Herman Daly's (1991) *Steady-State Economics,* Erik Olin Wright's (2010) *Envisioning Real Utopias,* Charles Eisenstein's (2011) *Sacred Economics: Money, Gift and Society in the Age of Transition,* David Schweickart's (2011) *After Capitalism,* or the NEF "Happy Planet Index" (2013).

Chapter 5

1. Environmental justice and ecofeminist scholars offer different frameworks and terminology for a radical sustainability (Pellow 2018; Gaard 2017). Some term this "just sustainabilities" (Agyeman, Bullard, and Evans 2003), "critical environmental justice" (Pellow 2017), or "just ecofeminist sustainability" (Gaard 2017), to name a few. Their overlapping concerns are ecological restoration and social equity. Thus, regenerative and transformative movements offer ecological restoration and liberating social justice. I go into more detail about each in the first chapter.
2. Of course, there are examples of traditional or Indigenous communities altering ecologies and exploiting natural resources. One example is how Native American communities' affected abalone in the area now known as California (Erlandson, Rick, and Braje 2009; Longo, Clausen, and Clark 2015). Diamond writes extensively about the Mayan civilization collapse as well as others (Diamond 2005).
3. I analyze interview data, their welcome pamphlet, and villagers' everyday actions to distinguish ecovillagers' goals and values. First, I identify recurring themes from

interviewees' responses to questions regarding ecovillage community values, personal values, their understanding of dominant cultural values, and problems they see both in the ecovillage and in dominant culture. Then I evaluate how this understanding of values translates into everyday action by examining responses individuals gave to questions regarding what a typical day looks like, what a typical day might look like if they did not live at the ecovillage, and how individuals feel the ecovillage is affected by its location in a city. Finally, I use my observations in my field journal to confirm, disconfirm, and contextualize interviewees' responses to my questions.

4. I do not mean to romanticize traditional communities. The communities of the Mediterranean in particular have unequal gender relations that are not ideal (Longo, Clausen, and Clark 2015).

Conclusion

1. Constructing south facing windows is a passive solar building technique designed to bring in more sunlight and absorb the sun's heat during winter months.
2. These oil and gas extraction processes are happening to preserve our fossil-fuel based economy, which was, until recently (Lazard 2018), more economically efficient than renewable forms of energy, though is still politically and infrastructurally entrenched as the dominant global energy source (EPA 2018; US Energy Information Administration 2019). Ralph's concern is that even if climate change doesn't kill us first, the energy and material production transition that is needed to stave off climate change will devastate our economy. Though, ecological and environmental economists argue that if we do nothing to stop climate change, the current and long-term economic costs will far outweigh any current or past savings (NRDC 2008).
3. Carbon emissions have risen in both 2017 and 2018, despite having plateaued since the 2007–2008 recession (Global Carbon Project 2018). However, in 2020, mainly as a result of the effects of the Covid-19 pandemic on the global economy, emissions fell (IEA 2021).

References

Achbar, Mark, and Jennifer Abbott. 2003. *The Corporation: The Pathological Pursuit of Profit and Power.* Toronto, Canada: Big Picture Media.

Acker, Joan. 2006. *Class Questions: Feminist Answers.* New York: Rowman and Littlefield Publishers.

Adams, Matthew, and Martin Jordan. 2016. "Growing Together: Nature Connectedness, Belonging and Social Identity in a Mental Health Ecotherapy Program." In *Ecotherapy: Theory, Research, and Practice,* edited by Martin Jordan and Joe Hinds, 122–36. New York: Palgrave.

Agyeman, Julian, Robert Bullard, and B. Evans. 2003. "Joined-Up Thinking: Bringing Together Sustainability, Environmental Justice and Equity." In *Just Sustainabilities: Development in an Unequal World,* edited by Julian Agyeman, Robert Bullard, and B. Evans, 1–16. London: Earthscan.

Alexander, Bruce. 2008. *The Globalization of Addiction: A Study in the Poverty of the Spirit.* New York: Oxford University Press.

Alexander, Bruce. 2010. "Addiction: The View from Rat Park." Accessed September 29, 2018. http://www.brucekalexander.com/articles-speeches/rat-park/148-addiction-the-view-from-rat-park.

Alexander, Michelle. 2010. *The New Jim Crow: Mass Incarceration in the Age of Colorblindness.* New York: New Press.

Alkon, Allison Hope, and Julian Agyeman. 2011. *Cultivating Food Justice: Race, Class and Sustainability.* Cambridge, MA: MIT Press.

Altieri, Miguel. 2009. "Agroecology, Small Farms, and Food Sovereignty." *Monthly Review* 61 (3). https://monthlyreview.org/2009/07/01/agroecology-small-farms-and-food-sovereignty/.

Altieri, Miguel. 2017. "Miguel Altieri: Professor of Department of Environmental Science, Policy, and Management, Berkeley." UC Berkeley Department of Environmental Science, Policy, and Management website. Accessed December 23, 2017. https://ourenvironment.berkeley.edu/people/miguel-altieri.

Altieri, Miguel, and Fernando R. Funes-Monzote. 2012. "The Paradox of Cuban Agriculture." *Monthly Review* 63 (8). https://ourenvironment.berkeley.edu/people/miguel-altieri.

Alvarez, Camila. 2016. "Militarization and Water: A Cross-National Analysis of Militarism and Freshwater Withdrawals." *Environmental Sociology* 2 (3): 298–305.

Amazon Environmental Research Institute, the Institute for Health Policy Studies, and Human Rights Watch. 2020. "'The Air is Unbearable': Health Impacts of Deforestation-Related Fires in the Brazilian Amazon." Accessed October 7, 2020. https://www.hrw.org/report/2020/08/26/air-unbearable/health-impacts-deforestation-related-fires-brazilian-amazon.

American Civil Liberties Union. 2020. "Family Separation: By the Numbers" Accessed October 7, 2020. https://www.aclu.org/issues/immigrants-rights/immigrants-rights-and-detention/family-separation.

American Psychological Association. 2004. "Driving Teen Egos—and Buying—through 'Branding' a Glut of Marketing Messages Encourages Teens to Tie Brand Choices to Their Personal Identity." Accessed October 5, 2017. http://www.apa.org/monitor/jun04/driving.aspx.

American Psychiatric Association. 2013. *Diagnostic and Statistical Manual of Mental Disorders*. 5th ed. DSM-5. https://doi.org/10.1176/appi.books.9780890425596

Anda, Robert F., David W. Brown, Vincent J. Felitti, J. Douglas. Bremner, Shanta R. Dube, and Wayne H. Giles. 2007. "Adverse Childhood Experiences and Prescribed Psychotropic Medications in Adults." *American Journal of Preventive Medicine* 32 (5): 389–94.

Anda, Robert F., David W. Brown, Vincent J. Felitti, Shanta R. Dube, and Wayne H. Giles. 2008. "Adverse Childhood Experiences and Prescription Drug Use in a Cohort Study of Adult HMO Patients." *BMC Public Health* 8 (198). doi: 10.1186/1471-2458-8-198.

Anda, Robert F., Maxia Dong, David W. Brown, Vincent J. Felitti, Wayne H. Giles, Geraldine S. Perry, Edwards J. Valerie, and Shanta R. Dube. 2009. "The Relationship of Adverse Childhood Experiences to a History of Premature Death of Family Members." *BMC Public Health* 9 (106). doi: 10.1186/1471-2458-9-106.

Anderson, Charles H. 1976. *The Sociology of Survival: Social Problems of Growth*. Homewood, IL: Dorsey Press.

Anderson, Kat. 2005. *Tending the Wild: Native American Knowledge and the Management of California's Natural Resources*. Los Angeles: University of California Press.

Aptekar, Sofya. 2015. "Visions of Public Space: Reproducing and Resisting Social Hierarchies in a Community Garden." *Sociological Forum* 30 (1): 209–27.

Atalay, Atalay A., and Margaret G. Meloy. 2011. "Retail Therapy: A Strategic Effort to Improve Mood." *Psychology & Marketing* 28 (6): 638–60.

Bajekal, Naina. 2018. "Want to Win the War on Drugs? Portugal Might Have the Answer." *Time*. Accessed April 23, 2019. http://time.com/longform/portugal-drug-use-decriminalization/.

Barbosa, Luiz. 2015. "Theories in Environmental Sociology." In *Twenty Lessons in Environmental Sociology*, edited by Kenneth Gould and Tammy Lewis, 28–50. New York: Oxford University Press.

Barnosky, Anthony D., James H. Brown, Gretchen C. Daily, Rodolfo Dirzo, Anne H. Ehrlich, Paul R. Ehrlich, Jussi T. Eronen, et al. 2014. "Introducing the Scientific Consensus on Maintaining Humanity's Life Support Systems in the 21st Century: Information for Policy Makers." *The Anthropocene Review* 1 (1): 78–109.

BBC. 2017. "Cuba Profile—Timeline." *BBC News*. Accessed December 21, 2017. http://www.bbc.com/news/world-latin-america-19576144.

Beaubien, Jason. 2017. "Unhappy Anniversary, South Sudan." *NPR*. Accessed September 7, 2017. http://www.npr.org/sections/goatsandsoda/2017/07/08/535620121/unhappyanniversarysouth-sudan.

Bell, Shannon Elizabeth, and Richard York. 2010. "Community Economic Identity: The Coal Industry and Ideology Construction in West Virginia." *Rural Sociology* 75 (1): 111–43.

Bendell, Jem. 2018. "Deep Adaptation: A Map for Navigating Climate Tragedy." *IFLAS Occasional Paper 2*. Accessed July 9, 2019. www.iflas.info.

Berg, Bruce L. 2007. *Qualitative Research Methods for the Social Sciences, 6th ed.* England: Pearson Education.

Bernays, Edward. 1928. *Propaganda*. New York: Horace Liveright.

Bhada-Tata, Perinaz, and Daniel A. Hoornweg. 2012. "What a Waste? A Global Review of Solid Waste Management." *Urban Development Series Knowledge Papers*, no. 15. Washington, DC: World Bank Group. Accessed June 28, 2016. http://documents.worldbank.org/curated/en/2012/03/20213522/waste-global-review-solid-waste-management.

Bhattacharya, Tithi. 2017. *Social Reproduction Theory: Remapping Class, Recentering Oppression*. London, UK: Pluto Press.

Biello, David. 2009. "Can Climate Change Cause Conflict? Recent History Suggests So." *Scientific American*. Accessed September 7, 2017. https://www.scientificamerican.com/article/can-climate-change-cause-conflict/.

Blue Cross Blue Shield. 2017. "America's Opioid Epidemic and Its Effect on the Nation's Commercially-Insured Population." The Health of America. Accessed on October 25, 2018. https://www.bcbs.com/the-health-of-america/reports/americas-opioid-epidemic-and-its-effect-on-the-nations-commercially-insured.

Bockman, Johanna. 2013. "Neoliberalism." *Contexts* 12 (3): 14–15.

Boggs, Carl, Jr. 1977. "Revolutionary Process, Political Strategy, and the Dilemma of Power." *Theory and Society* 4 (3): 359–93.

Boyce, James K. 2008. "Is Inequality Bad for the Environment?" *Research in Social Problems and Public Policy* 15:267–88.

Boyd, Emily. 2002. "The Noel Kempff Project in Bolivia: Gender, Power, and Decision-Making in Climate Mitigation." *Gender and Development* 10 (2): 70–77.

Brand, Peter, and Michael J. Thomas. 2005. *Urban Environmentalism: Global Change and the Mediation of Local Conflict*. New York: Routledge.

Braun, Yvonne A. 2008. "'How Can I Stay Silent?' One Woman's Struggle for Environmental Justice in Lesotho." *Journal of International Women's Studies* 10 (1): 5–20.

Brittle, Zach. 2014. "D Is for Defensiveness." The Gottman Institute. Accessed July 13, 2019. https://www.gottman.com/blog/d-is-for-defensiveness/.

Broom, Douglas. 2019. "The Dirty Secret of Electric Vehicles." World Economic Forum. Accessed August 2, 2019. https://www.weforum.org/agenda/2019/03/the-dirty-secret-of-electric-vehicles/.

Broswimmer, Franz. 2002. *Ecocide: A Short History of the Mass Extinction of Species*. Ann Arbor, MI: Pluto Press.

Brown, Adrienne Maree. 2019. *Pleasure Activism: The Politics of Feeling Good*. Chico, CA: AK Press.

Brown, Autumn, and adrienne maree brown. 2019. *How to Survive the End of the World*, produced by Zak Rosen. Podcast audio. Accessed August 1, 2019. https://www.endoftheworldshow.org/.

Brown, David W., Robert F. Anda, Vincent J. Felitti, Valerie J. Edwards, Ann Marie Malarcher, Janet B. Croft, and Wayne H. Giles. 2010. "Adverse Childhood Experiences and the Risk of Lung Cancer." *BMC Public Health* 10 (20). https://doi.org/10-1186/1471-2458-10-20.

Buck, Daniel, Christina Getz, and Julie Guthman. 1997. "From Farm to Table: The Organic Vegetable Commodity Chain of Northern California." *Sociologia Ruralis* 37 (1): 3–20.

Buckingham, Susan. 2010. "Call in the Women." *Nature* 468 (502).

Burawoy, Michael. 1998. "The Extended Case Method." *Sociological Theory* 16 (1): 4–33.

Burgess, Rod, Marisa Carmona, and Theo Kolstee. 1997. "Cities, the State and the Market." In *The Challenge of Sustainable Cities: Neoliberalism and Urban Strategies in Developing Countries,* edited by Rod Burgess, Marisa Carmona, and Theo Kolstee, 3–16. Atlantic Highlands, NJ: Zed Books.

Burkhead, Noel M. 2012. "Extinction Rates in North American Freshwater Fishes, 1900–2010." *BioScience* 62 (9): 798–808.

Burrow, Sylvia. 2005. "The Political Structure of Emotion: From Dismissal to Dialogue." *Hypatia* 20 (4): 27–43.

Cacioppo, John, and William Patrick. 2008. *Loneliness: Human Nature and the Need for Social Connection.* New York: W.W. Norton.

Cal Fire. 2020. "California Statewide Fire Summary October 6, 2020." California Daily Wildfire Update. Accessed October 7, 2020. https://www.fire.ca.gov/daily-wildfire-report/.

Callero, Peter. 2018. *The Myth of Individualism: How Social Forces Shape Our Lives.* New York: Rowman and Littlefield.

Cama, Timothy. 2015. "Sanders: Paris Climate Pact 'Goes Nowhere Near far Enough.'" *The Hill.* Accessed September 9, 2017. http://thehill.com/policy/energy-environment/263042-sanders paris-climate-pact-goes-nowhere-near-far-enough.

Campbell, B. J. 2018. "The Surprisingly Solid Mathematical Case of the Tin Foil Hat Gun Prepper: Or, 'Who Needs an AR-15 Anyway?'" *Medium.* Accessed June 28, 2019. https://medium.com/s/story/the-surprisingly-solid-mathematical-case-of-the-tin-foil-hat-gun-prepper-15fce7d10437.

Čapek, Stella M. 2010. "Foregrounding Nature: An Invitation to Think about Shifting Nature-City Boundaries." *City and Community* 9 (2): 208–24.

Cardinale, Bradley J., J. Emmett Duffy, Andrew Gonzalez, David U. Hooper, Charles Perrings, Patrick Venail, Anita Narwani, et al. 2012. "Biodiversity Loss and Its Impact on Humanity." *Nature* 486 (7401): 59–67.

Carfagna, Lindsey, Emilie A. Dubois, Connor Fitzmaurice, Monique Y. Ouimette, Juliet B. Schor, and Margaret Willis. 2014. "An Emerging Eco-Habitus: The Reconfiguration of High Cultural Capital Practices among Ethical Consumers." *Journal of Consumer Culture* 14 (2): 158–78.

Carruyo, Light. 2008. *Producing Knowledge, Protecting Forests: Rural Encounters with Gender, Ecotourism, and International Aid in the Dominican Republic.* University Park: Pennsylvania State University Press.

Carson, Rachel. 1962. *Silent Spring.* Boston, MA: Houghton Mifflin.

Catton, William, and Riley Dunlap. 1978. "Environmental Sociology: A New Paradigm." *The American Sociologist* 13 (1): 41–49.

Ceballos, Gerardo, Paul R. Ehrlich, and Rodolfo Dirzo. 2017. "Biological Annihilation Via the Ongoing Sixth Mass Extinction Signaled by Vertebrate Population Losses and Declines." *PNAS* 114 (30): E6089–E6096.

Center for Climate and Security. 2017. "About." Accessed September 9, 2017. https://climateandsecurity.org/about/.

Center for Sustainable Economy. 2019. "Measures of Progress." Accessed August 20, 2019. https://sustainable-economy.org/new-measures-progress/.

Chapman, Daniel P., Robert F. Anda, Vincent J. Felitti, Shanta R. Dube, Valerie J. Edwards, and Charles L. Whitfield. 2004. "Adverse Childhood Experiences and the Risk of Depressive Disorders in Adulthood." *Journal of Affective Disorders* 82 (2): 217–25.

Chapman, Daniel P., Shanta R. Dube, and Robert F. Anda. 2007. "Adverse Childhood Events as Risk Factors for Negative Mental Health Outcomes." *Psychiatric Annals* 37 (5): 359–64.

Checker, Melissa. 2011. "Wiped Out by the 'Greenwave': Environmental Gentrification and the Paradoxical Politics of Urban Sustainability." *City and Society* 23 (2): 210–29.

Cigna. 2018. "Cigna 2018 US Loneliness Index: Survey of 20,000 Americans Examining Behaviors Driving Loneliness in the United States." Accessed October 28, 2019. https://www.cigna.com/about-us/newsroom/studies-and-reports/loneliness-epidemic-america.

Cillizza, Chris. 2017. "Donald Trump Doesn't Think Much of Climate Change, in 20 Quotes." *CNN*. Accessed October 30, 2017. http://www.cnn.com/2017/08/08/politics/trump-global-warming/index.html.

Clark, Brett, and Richard York. 2005. "Carbon Metabolism: Global Capitalism, Climate Change, and the Biospheric Rift." *Theory and Society* 34 (4): 391–428.

Clarke, Lee, and Caron Chess. 2008. "Elites and Panic: More to Fear than Fear Itself." *Social Forces*. 87 (2): 993–1014.

Clausen, Rebecca. 2009. "Healing the Rift: Metabolic Restoration in Cuban Agriculture." In *Environmental Sociology: From Analysis to Action*, edited by L. King and D. McCarthy, 425–37. New York: Rowman and Littlefield Publishers.

Clayton, Susan Whitmore-Williams, Christie M. Manning, Kirra Krygsman, and Meighen Speiser. 2017. *Mental Health and Our Changing Climate: Impacts, Implications, and Guidance*. Washington, DC: American Psychological Association and ecoAmerica.

Clement, Matthew Thomas. 2010. "Urbanization and the Natural Environment: An Environmental Sociological Review and Synthesis." *Organization & Environment* 23 (3): 291–314.

Clement, Matthew Thomas. 2011a. "'Let Them Build Sea Walls': Ecological Crisis, Economic Crisis and the PEOS." *Critical Sociology* 37 (4): 447–63.

Clement, Matthew Thomas. 2011b. "The Town-Country Antithesis and the Environment: A Sociological Critique of a 'Real Utopian' Project." *Organization & Environment* 24 (3): 292–311.

Clement, Matthew, Christina Ergas, and Patrick Greiner. 2015. "The Environmental Consequences of Rural and Urban Population Change: An Exploratory Spatial Panel Study of Forest Cover in the Southern United States, 2001–2006." *Rural Sociology* 80 (1): 108–36.

Cohen, Sheldon, and Denise Janicki-Deverts. 2012. "Who's Stressed? Distributions of Psychological Stress in the United States in Probability Samples from 1983, 2006, and 2009." *Journal of Applied Social Psychology* 42 (6): 1320–34.

Colantonio, Andrea, and Robert Potter. 2006. "The Rise of Urban Tourism in Havana Since 1989." *Geography* 91 (1): 23–33.

Collins, Chuck, and Josh Hoxie. 2017. *Billionaire Bonanza: The Forbes 400 and the Rest of Us*. Washington, DC: The Institute for Policy Studies.

Collins, Patricia Hill. 2000. *Black Feminist Thought: Knowledge, Consciousness, and the Politics of Empowerment*. New York: Routledge.

Colten, Craig. 2005. *An Unnatural Metropolis: Wresting New Orleans from Nature*. Baton Rouge: Louisiana State University Press.

Commoner, Barry. 1992. *Making Peace with the Planet*. New York: New Press.

Companioni, Nelso, Yanet Ojeda Hernández, Egidio Páez, and Catherine Murphy. 2002. "The Growth of Urban Agriculture." In *Sustainable Agriculture and*

Resistance: Transforming Food Production in Cuba*, edited by Fernando Funes, Luis García, Martin Bourque, Nilda Pérez, and Peter Rosset, 220–36. Oakland, CA: Food First Books.

Conover, P. 1975. "An Analysis of Communes and Intentional Communities with Particular Attentions to Sexual and Gender Relations." *The Family Coordinator* 24: 453–64.

Cook, John, Naomi Oreskes, Peter T. Doran, William R. L. Anderegg, Bart Verheggen, Ed W. Maibach, J. Stuart Carlton, et al. 2016. "Consensus on Consensus: A Synthesis of Consensus Estimates on Human-Caused Global Warming." *Environmental Research Letters* 11(4). https://doi.org/10.1088/1748-9326/11/4/048002.

Coyle, Kevin, and Lise Van Susteren. 2011. "The Psychological Effects of Global Warming on the United States: And Why the U.S. Mental Health Care System Is Not Adequately Prepared." *National Forum Research Report*. Merrifield, VA: National Wildlife Federation Climate Education Program.

Crenshaw, Kimberle. 1991. "Mapping the Margins: Intersectionality, Identity Politics, and Violence against Women of Color." *Stanford Law Review* 43 (6): 1241–99.

Cronen, William. 1991. *Nature's Metropolis: Chicago and the Great West*. New York: W. W. Norton.

Crowley, Thomas J. 2000. "Causes of Climate Change over the Past 1000 Years." *Science* 289 (5477): 270–77.

Crutzen, Paul J., and Christian Schwägerl. 2011. "Living in the Anthropocene: Toward a New Global Ethos." *Yale Environment* 360.

Curtin, Sally C., Margaret Warner, and Holly Hedegaard. 2016. "Increase in Suicide in the United States, 1999–2014." Center for Disease Control, NCHS Data Brief no. 241. Accessed October 3, 2018. https://www.cdc.gov/nchs/products/databriefs/db241.htm.

Cushing, Lara, Rachel Morello-Frosch, Madeline Wander, and Manuel Pastor. 2015. "The Haves, the Have-Nots, and the Health of Everyone: The Relationship between Social Inequality and Environmental Quality." *Annual Review Public Health* 36: 193–209. https://doi.org/10.1146/annurev-publhealth-031914-122646.

Daly, Herman. 1991. *Steady-State Economics*. 2nd ed. Washington, DC: Island Press.

Daly, Herman. 1996. *Beyond Growth: The Economics of Sustainable Development*. Boston, MA: Beacon Press.

Davenport, Coral. 2015. "Nations Approve Landmark Climate Accord in Paris." *The New York Times*. Accessed September 9, 2017. https://www.nytimes.com/2015/12/13/world/europe/climate-change-accord-paris.html.

Davenport, Coral, and Alissa J. Rubin. 2017. "Trump Signs Executive Order Unwinding Obama Climate Policies" *The New York Times*. Accessed April 18, 2017. https://www.nytimes.com/2017/03/28/climate/trump-executive-order-climate-change.html.

Davis, Mike. 1990. *City of Quartz: Excavating the Future in Los Angeles*. New York: Verso.

Davis, Mike. 2006. *Planet of Slums*. New York: Verso.

Davis, Angelique M., and Rose Ernst. 2017. "Racial Gaslighting." *Politics, Groups, and Identities* 5. https://doi.org/10.1080/21565503.2017.1403934.

Dawson, Michael. 2003. *The Consumer Trap: Big Business Marketing in American Life*. Urbana: University of Illinois Press.

DeAngelis, Tori. 2004. "Consumerism and Its Discontents: Materialistic Values May Stem from Early Insecurities and Are Linked to Lower Life Satisfaction, Psychologists Find. Accruing More Wealth May Provide Only a Partial Fix." *American Psychological*

Association 35 (6): 52. Accessed Oct. 4, 2017. http://www.apa.org/monitor/jun04/discontents.aspx.

Diamond, Jared. 1987. "The Worst Mistake in the History of the Human Race." *Discover Magazine*. Accessed December 19, 2017. http://discovermagazine.com/1987/may/02-the-worst-mistake-in-the-history-of-the-human-race.

Diamond, Jared. 2005. *Collapse: How Societies Choose to Fail or Succeed*. New York: Penguin.

Dietz, Thomas, Eugene A. Rosa, and Richard York. 2007. "Driving the Human Ecological Footprint." *Frontiers in Ecology and the Environment* 5 (1): 13–18.

Dillard, Jennifer. 2008. "A Slaughterhouse Nightmare: Psychological Harm Suffered by Slaughterhouse Employees and the Possibility of Redress through Legal Reform." *Georgetown Journal on Poverty Law & Policy* 15 (2): 391–408.

Dillard, Jesse, Veronica Dujon, and Mary C. King. 2009. "Introduction." In *Understanding the Social Dimensions of Sustainability*, edited by Jesse Dillard, Veronica Dujon, and Mary C. King, 1–12. New York: Routledge.

Dockrill, Peter. 2018. "An MIT Computer Predicted the End of Civilization Almost 50 Years Ago." *Science Alert*. Accessed July 13, 2019. https://www.sciencealert.com/how-mit-computer-predicted-end-civilisation-almost-50-years-ago-world1-world3-club-rome-limits-growth.

Doherty, Thomas. 2016. "Theoretical and Empirical Foundations for Ecotherapy." In *Ecotherapy: Theory, Research, and Practice*, edited by Martin Jordan and Joe Hinds, 12–31. New York: Palgrave.

Domhoff, G. William. 2009. *Who Rules America? Challenges to Corporate and Class Dominance*. New York: McGraw-Hill.

Dong, Ensheng, Hongru Du, Lauren Gardner. 2020. An Interactive Web-Based Dashboard to Track COVID-19 in Real Time. *The Lancet Infectious Diseases* 20 (5): 533–34. doi: 10.1016/S1473-3099(20)30120-1. Accessed October 7, 2020. https://www.arcgis.com/apps/opsdashboard/index.html#/bda7594740fd40299423467b48e9ecf6.

Downey, Liam. 2015. *Inequality, Democracy, and the Environment*. New York: New York University Press.

Downey, Liam, Eric Bonds, and Katherine Clark. 2010. "Natural Resource Extraction, Armed Violence, and Environmental Degradation." *Organization & Environment* 23 (4): 417–45.

Dube, Shanta R., DeLisa Fairweather, William S. Pearson, Vincent J. Felitti, Robert F. Anda, and Janet B. Croft. 2009. "Cumulative Childhood Stress and Autoimmune Disease." *Psychomatic Medicine* 71 (2): 243–50.

Dube, Shanta R., Vincent J. Felitti, Maxia Dong, Daniel P. Chapman, Wayne H. Giles, and Robert F. Anda. 2003. "Childhood Abuse, Neglect, and Household Dysfunction and the Risk of Illicit Drug Use: The Adverse Childhood Experiences Study." *Pediatrics* 111 (3): 564–72.

Dube, Shanta R., Jacqueline W. Miller, David W. Brown, Wayne H. Giles, Vincent J. Felitti, Maxia Dong, and Robert F. Anda. 2006. "Adverse Childhood Experiences and the Association with Ever Using Alcohol and Initiating Alcohol Use During Adolescence." *Journal of Adolescent Health* 38 (4): 444.e1–444.e10.

Dunaway, Wilma. 2014. "Bringing Commodity Chain Analysis Back to Its World-Systems Roots: Rediscovering Women's Work and Households." *American Sociological Association* 20 (1): 64–81.

"East India Company, British." 2018. In *The Columbia Encyclopedia*, edited by Paul Lagasse and Columbia University. New York: Columbia University Press. Credo Reference. Accessed October 25, 2018. http://proxy.lib.utk.edu:90/login?url=https://search.credoreference.com/content/entry/columency/east_india_company_british/0?institutionId=680.

Eckhardt, Giana M., and Fleura Bardhi. 2015. "The Sharing Economy Isn't about Sharing at All." *Harvard Business Review*. Accessed October 15, 2017. https://hbr.org/2015/01/the-sharing-economy-isnt-about-sharing-at-all.

Eckstein, Barbara, and James A. Throgmorton. 2003. *Story and Sustainability: Planning, Practice, and Possibility for American Cities*. Cambridge, MA: MIT Press.

Economic Freedom Index. 2013. "Interactive Heatmap." Index of Economic Freedom. Accessed October 18, 2013. http://www.heritage.org/index/heatmap.

Economist Intelligence Unit. 2012. "Democracy Index 2012: Democracy at a Standstill." Accessed October 18, 2013. http://www.eiu.com/Handlers/WhitepaperHandler.ashx?fi=Democracy-Index- 2012.pdf&mode=wp&campaignid=DemocracyIndex12.

Edwards, Valerie J., Robert F. Anda, David Gu, Shanta R. Dube, and Vincent J. Felitti. 2007. "Adverse Childhood Experiences and Smoking Persistence in Adults with Smoking-Related Symptoms and Illness." *Permanente Journal* 11, 2: 5–13.

Egan, Elisabeth. 2020. "Timing, Patience and Wisdom Are the Secrets to Robin Wall Kimmerer's Success." *The New York Times*. Accessed November 10, 2020. https://www.nytimes.com/2020/11/05/books/review/robin-wall-kimmerer-braiding-sweetgrass.html?smid=tw-nytbooks&smtyp=cur

Ehrenreich, Barbara. 1976. "What Is Socialist Feminism?" Reprinted in *Working Papers on Socialism and Feminism*. Chicago: New American Movement.

Ehrenreich, Barbara. 2009. *Bright-Sided: How Positive Thinking Is Undermining America*. New York: Picador.

Ehrlich, Paul, and John Holdren. 1971. "Impact of Population Growth." *Science* 171: 1212–17.

Eisenstein, Charles. 2011. *Sacred Economics: Money, Gift and Society in the Age of Transition*. Berkeley, CA: North Atlantic Books.

E-Marketer. 2014. "Advertisers Will Spend Nearly $600 Billion Worldwide in 2015: US, China, Japan, Germany and the UK Lead as the Top Five Ad Markets." Accessed April 17, 2017. https://www.emarketer.com/Article/Advertisers-Will-Spend-Nearly-600-Billion-Worldwide-2015/1011691.

Emerson, Robert M., Rachel I. Fretz, and Linda L. Shaw. 1995. *Writing Ethnographic Fieldnotes*." Chicago: University Press of Chicago.

Environmental Protection Agency. 2013. "Advancing Sustainable Materials Management: Facts and Figures Report." Accessed June 28, 2016. https://www.epa.gov/smm/advancing-sustainable-materials-management-facts-and-figures.

Environmental Protection Agency. 2016a. "Green Building." Accessed November 11, 2017. https://archive.epa.gov/greenbuilding/web/html/.

Environmental Protection Agency. 2016b. "Learn about Sustainability: What Is Sustainability?" Accessed October 30, 2017. https://www.epa.gov/sustainability/learn-about-sustainability#what.

Environmental Protection Agency. 2017. "Global Greenhouse Gas Emissions Data." Accessed September 9, 2017. https://www.epa.gov/ghgemissions/global-greenhouse-gas-emissions-data.

Environmental Protection Agency. 2018. "The Process of Unconventional Natural Gas Production: Hydraulic Fracturing." Accessed July 1, 2019. https://www.epa.gov/uog/process-unconventional-natural-gas-production.

Ergas, C. 2010. "A Model of Sustainable Living: Collective Identity in an Urban Ecovillage." *Organization & Environment* 23 (1): 32–54.

Ergas, C. 2013a. "Barriers to Sustainability: A Qualitative Cross-National Comparison." PhD diss., University of Oregon.

Ergas, C. 2013b. "Cuban Urban Agriculture as a Strategy for Food Sovereignty." Review of *Sustainable Urban Agriculture in Cuba*, by Sinan Koont. *Monthly Review* 64 (10): 46–52.

Ergas, C. 2014. "Barriers to Sustainability: Gendered Divisions of Labor in Cuban Urban Agriculture." In *From Sustainable to Resilient Cities: Global Concerns and Urban Efforts*, vol. 14, edited by William G. Holt. Bingley, UK: Emerald Group Publishing.

Ergas, Christina, and Matthew Clement. 2016. "Ecovillages, Restitution, and the Political-Economic Opportunity Structure: An Urban Case Study in Mitigating the Metabolic Rift." *Critical Sociology* 42 (7–8): 1195–1211.

Ergas, Christina, Matthew Clement, and Julius McGee. 2016. "Urban Density and the Metabolic Reach of Metropolitan Areas: A Panel Analysis of Per Capita Transportation Emissions at the County Level." *Social Science Research* 58: 243–53.

Ergas, Christina, and Richard York. 2012. "Women's Status and Carbon Dioxide Emissions: A Quantitative Cross-National Analysis." *Social Science Research* 41: 965–76.

Erlandson, Jon M., Torben C. Rick, and Todd J. Braje. 2009. "Fishing up the Food Web? 12,000 Years of Maritime Subsistence and Adaptive Adjustments on California's Channel Islands." *Pacific Science* 63 (4): 711–24.

European Economic and Social Committee. 2018. "Fascism On the Rise: Where Does It Come From, and How to Stop It, With a Common European Response." Accessed October 7, 2020. https://www.eesc.europa.eu/en/news-media/news/fascism-rise-where-does-it-come-and-how-stop-it-common-european-response.

European Monitoring Centre for Drugs and Drug Addiction. 2015. *European Drug Report: Trends and Developments*. Luxembourg: Publications Office of the European Union.

European Monitoring Centre for Drugs and Drug Addiction. 2018. *Portugal: Country Drug Report 2018*. Luxembourg: Publications Office of the European Union. www.emcdda.europa.eu/countries/drug-reports/2018/portugal_en.

Farrell, Justin. 2015. "Network Structure and Influence of the Climate Change Counter-Movement." *Nature Climate Change* 6. https://doi.org/10.1038/NCLIMATE2875.

Fellowship of Intentional Communities. 2017. "Communities by Country." Accessed July 25, 2017. http://www.ic.org/the-fellowship-for-intentional-community/.

Fitzgerald, Amy J., Linda Kalof, and Thomas Dietz. 2009. "Slaughterhouses and Increased Crime Rates: An Empirical Analysis of the Spillover from 'The Jungle' into the Surrounding Community." *Organization Environment* 22 (2): 158.

Flint, Warren. 2013. *Practice of Sustainable Community Development: A Participatory Framework for Change*. New York: Springer.

Foster, John Bellamy. 1999. "Marx's Theory of Metabolic Rift: Classical Foundations for Environmental Sociology." *American Journal of Sociology* 105 (2): 366–405.

Foster, John Bellamy. 2000. *Marx's Ecology: Materialism and Nature*. New York: Monthly Review Press.

Foster, John Bellamy. 2008. *The Sustainability Mirage: Illusion and Reality in the Coming War on Climate Change*. Sterling, VA: Earthscan.

Foster, John B., Brett Clark, and Richard York. 2010. *The Ecological Rift: Capitalism's War on the Earth*. New York: Monthly Review Press.

Freedom House. 2017. "Freedom in the World 2017." Accessed December 22, 2017. https://freedomhouse.org/report/freedom-world/freedom-world-2017?gclid=EAIaIQobChMI___O6pme2AIVUrnACh04uwH7EAAYASAAEgLY__D_BwE.

French, Bryana, Jioni Lewis, Della V. Mosley, Hector Y. Adames, Nayeli Y. Chavez-Dueñas, Grace A. Chen, and Helen A. Neville. 2019. "Toward a Psychological Framework of Radical Healing in Communities of Color." *The Counseling Psychologist*: 1–33.

Friedman, Thomas L. 2008. *Hot, Flat, and Crowded: Why We Need A Green Revolution and How it Can Renew America*. New York: Farrar, Straus, and Giroux.

Fromm, Erich. 1956. *The Art of Loving*. New York: Harper Collins Publishers

Gaard, Greta. 2017. *Critical Ecofeminism*. New York: Lexington Books.

Garber, Kent. 2008. "Behind the Prosperity Gospel: Followers Believe God Wants Them to Be Rich—Not Just Spiritually but Materially." *U.S. News and World Report*. Accessed October 9, 2017. https://www.usnews.com/news/national/articles/2008/02/15/behind-the-prosperity-gospel.

Gattinger, Andreas, Adrian Mullera, Matthias Haenia, Colin Skinnera, Andreas Fliessbacha, Nina Buchmannb, Paul Mädera, et al. 2012. "Enhanced Top Soil Carbon Stocks under Organic Farming." *PNAS* 109 (44): 18226–31.

Gilbert, Natasha. 2012. "One-Third of Our Greenhouse Gas Emissions Come from Agriculture: Farmers Advised to Abandon Vulnerable Crops in Face of Climate Change." *Nature*. Accessed November 7, 2017. https://www.nature.com/news/one-third-of-our-greenhouse-gas-emissions-come-from-agriculture-1.11708.

Gilman, Robert. 1991. "The Eco-Village Challenge: The Challenge of Developing a Community Living in Balanced Harmony—with Itself as well as Nature—Is Tough, but Attainable." *In Context* 29: 10–14.

Glaeser, Edward. 2011. *Triumph of the City: How Our Greatest Invention Makes Us Richer, Smarter, Greener, Healthier, and Happier*. New York: Penguin.

Gleick, Peter. 2014. "Water, Drought, Climate Change, and Conflict in Syria." *Weather, Climate, and Society* 6 (3): 331–40.

Global Carbon Project. 2018. "Global Carbon Budget 2018: An Annual Update of the Global Carbon Budget and Trends." Accessed July 11, 2019. https://www.globalcarbonproject.org/carbonbudget/.

Global Footprint Network. 2015. "Cloughjordan Leads the Way Toward Sustainable Living." Accessed November 11, 2017. https://www.footprintnetwork.org/2015/02/05/cloughjordan-ecovillage-leads-way-toward-sustainable-living-ireland/.

Gómez, Maria Victoria, and Marjo Kuronen. 2011. "Comparing Local Strategies and Practices: Recollections from Two Qualitative Cross-national Research Projects." *Qualitative Research* 11 (6): 683–97.

Good Grief Network. "10-steps to Personal Resilience and Empowerment in a chaotic climate." https://www.goodgriefnetwork.org/.

Goodwin, Jeff, and James Jasper. 2015. *The Social Movements Reader: Cases and Concepts*. 3rd ed. Malden, MA: Wiley Blackwell.

Gore, Al. 2006. *An Inconvenient Truth*. Directed by Davis Guggenheim. Paramount Classics.

Gore, D'Angelo, Eugene Kiely, and Lori Robertson. 2016. "Trump on Climate Change." Factcheck. Accessed April 18, 2017. http://www.factcheck.org/2016/11/trump-on-climate-change/.

Gould, Kenneth A., David Naguib Pellow, and Allan Schnaiberg. 2008. *The Treadmill of Production: Injustice and Unsustainability in the Global Economy*. Boulder, CO: Paradigm Publishers.

Green, Hardy. 2010. *The Company Town: The Industrial Edens and Satanic Mills That Shaped the American Economy*. New York: Basic Books

Greshko, Michael, Laura Parker, and Brian Clark Howard. 2017. "A Running List of How Trump Is Changing the Environment." *National Geographic*. Accessed September 7, 2017. http://news.nationalgeographic.com/2017/03/how-trump-is-changing-science-environment/.

Guimaraes, Roberto. 2004. "The Political and Institutional Dilemmas of Sustainable Development." In *Handbook of Development Studies*, edited by Gedeon M. Mudacumura and M. Shamsul Haque, 447–64. FL: Taylor & Francis.

Guthman, Julie. 2004. *Agrarian Dreams: The Paradox of Organic Farming in California*. Berkeley: University of California Press.

Hails, Chris, Jonathon Loh, and Steven Goldfinger. 2006. "Living Planet Report." Accessed January 12, 2010. http://assets.panda.org/downloads/living_planet_report.pdf.

Hancox, Dan. 2013. "Spain's Communist Model Village." *The Guardian*. October 19, 2013. Accessed June 23, 2016. https://www.theguardian.com/world/2013/oct/20/marinaleda-spanish-communist-village-utopia.

Haraway, Donna. 2015. "Anthropocene, Capitalocene, Plantationocene, Chthulucene: Making Kin." *Environmental Humanities* 6: 159–65.

Harding, Sandra. 1991. *Whose Science? Whose Knowledge: Thinking from Women's Lives*. Ithaca, New York: Cornell University Press.

Harvey, David. 1982. *The Limits to Capital*. Chicago: University of Chicago Press.

Hauer, Mathew. 2017. "Migration Induced by Sea-level Rise Could Reshape the US Population Landscape." *Nature Climate Change Letters* 7: 321–27.

Hedva, Johanna. 2019. "Sick Woman Theory." *Mask Magazine*. Accessed July 13, 2019. http://www.maskmagazine.com/not-again/struggle/sick-woman-theory?fbclid=IwAR3Uz7LxupsGm4coiM69psUUIS2sSazOkyhaZpXbzjXsTQQvIxiocFXlBHo.

Herbst, Chris. 2011. "'Paradoxical' Decline? Another Look at the Relative Reduction in Female Happiness." *Journal of Economic Psychology* 32 (5): 773–88.

Herring, Horace. 2002. "The Quest for Arcadia: British Utopian Communities." *Organization & Environment* 15 (2): 202–8.

Hochschild, Arlie. Russell, and Anne Machung. 1990. *The Second Shift*. New York: Avon Books.

Holleman, Hannah. 2012. "Energy Policy and Environmental Possibilities: Biofuels and Key Protagonists of Ecological Change." *Rural Sociology* 77 (2): 280–307.

Holleman, Hannah. 2017. "De-Naturalizing Ecological Disaster: Colonialism, Racism and the Global Dust Bowl of the 1930s." *Journal of Peasant Studies* 44 (1): 234–60.

Holleman, Hannah. 2018. *Dust Bowls of Empire: Imperialism, Environmental Politics, and the Injustice of "Green" Capitalism*. New Haven, CT: Yale University Press.

Hollis-Brusky, Amanda. 2015. *Ideas with Consequences: The Federalist Society and the Conservative Counterrevolution*. New York: Oxford University Press.

Holmgren, David. 2004. *Permaculture: Principles and Pathways Beyond Sustainability*. Hepburn Springs, Victoria: Holmgren Design Services.

Holmstrom, Nancy. 2003. "The Socialist Feminist Project." *Monthly Review* 54 (10). https://monthlyreview.org/2003/03/01/the-socialist-feminist-project/.

Holt, Douglas B. 2014. "Why the Sustainable Economy Movement Hasn't Scaled: Toward a Strategy That Empowers Main Street." In *Sustainable Lifestyles and the Quest for Plentitude: Case Studies of the New Economy*, edited by Juliet Schor and Craig J. Thompson. New Haven, CT: Yale University Press.

hooks, bell. 2000. *All About Love: New Visions*. New York: Harper Perennial.

Hoornweg, Daniel, Perinaz Bhada-Tata, and Chris Kennedy. 2013. "Environment: Waste Production Must Peak This Century." *Nature* 502 (7473): 615–17.

Hrdy, Sarah Blaffer. 2009. *Mothers and Others: The Evolutionary Origins of Mutual Understanding*. Cambridge, MA: Harvard University Press.

Hsiang, Solomon M., Marshall Burke, and Edward Miguel. 2013. "Quantifying the Influence of Climate on Human Conflict." *Science* 341 (6151): 1235367.

Indigenous Action Media. 2014. "Accomplices Not Allies: Abolishing the Ally Industrial Complex, An Indigenous Perspective." Accessed July 1, 2019. https://neym.org/accomplices-not-allies-abolishing-ally-industrial-complex.

Inglehart, Ronald. 1977. *The Silent Revolution: Changing Values and Political Styles among Western Publics*. Princeton, NJ: Princeton University Press.

Intergovernmental Panel on Climate Change. 2013. *Climate Change 2013: The Physical Science Basis*. Accessed June 29, 2016. https://www.ipcc.ch/report/ar5/.

Intergovernmental Panel on Climate Change. 2018. "Summary for Policymakers." In *Global Warming of 1.5°C. An IPCC Special Report on the Impacts of Global Warming of 1.5°C Above Pre-Industrial Levels and Related Global Greenhouse Gas Emission Pathways, in the Context of Strengthening the Global Response to the Threat of Climate Change, Sustainable Development, and Efforts to Eradicate Poverty*, edited by V. Masson-Delmotte, P. Zhai, H. O. Pörtner, D. Roberts, J. Skea, P. R. Shukla, A. Pirani, et al. Geneva, Switzerland: World Meteorological Organization.

International Energy Agency (IEA). 2021. "Global Energy Review: CO_2 Emissions in 2020, Understanding the Impacts of Covid-19 on Global CO_2 Emissions." https://www.iea.org/articles/global-energy-review-co2-emissions-in-2020.

Islam, S. Nazrul. 2015. *Inequality and Environmental Sustainability*. United Nations Department of Economic and Social Affairs. Working Paper No. 145. Accessed June 1, 2018. http://www.un.org/en/development/desa/.

Jacques, Peter J., Riley E. Dunlap, and Mark Freeman. 2008. "The Organisation of Denial: Conservative Think Tanks and Environmental Skepticism." *Environmental Politics* 17 (3): 349–85.

Järvensivu, Paavo, Tero Toivanen, Tere Vadén, Ville Lähde, Antti Majava, and Jussi T. Eronen. 2018. *Global Sustainable Development Report 2019*. Group of Independent Scientists. Accessed July 13, 2019. https://bios.fi/bios-governance_of_economic_transition.pdf.

Jenkins, Martin. 2003. "Prospects for Biodiversity." *Science* 302 (5648): 1175–77.

Jevons, William Stanley. 2001. "Of the Economy of Fuel." *Organization & Environment* 14 (1): 99–104.

Jonas, Andrew E. G., and David Wilson. 1999. "The City as a Growth Machine: Critical Reflections Two Decades Later." In *The Urban Growth Machine: Critical Perspectives, Two Decades Later*, edited by Andrew E. G. Jonas and David Wilson, 3–18. Albany: State University of New York Press.

Jorgenson, Andrew, and Brett Clark. 2009. "The Economy, Military, and Ecologically Unequal Relationships in Comparative Perspective: A Panel Study of the Ecological Footprints of Nations, 1975–2000." *Social Problems* 56: 621–646.

Jorgenson, Andrew. K., Brett Clark, and Jeffrey Kentor. 2010. "Militarization and the Environment: A Panel Study of Carbon Dioxide Emissions and the Ecological Footprints of Nations, 1970–2000." *Global Environmental Politics* 10: 7–29.

Keeley, Lawrence. 1996. *War Before Civilization*. New York: Oxford University Press.

Kellogg, Scott, and Stacy Pettigrew. 2008. *Toolbox for Sustainable City Living: A Do-It-Ourselves Guide*. Boston: South End Press.

Kellstedt, Paul, Sammy Zahran, and Arnold Vedlitz. 2008. "Personal Efficacy, the Information Environment, and Attitudes Toward Global Warming and Climate Change in the United States." *Risk Analysis* 28 (1): 113–26.

Kemp, Rene, Johan Schot, and Remco Hoogma. 1998. "Regime Shifts to Sustainability Through Processes of Niche Formation: The Approach of Strategic Niche Management." *Technology Analysis and Strategic Management* 10 (2): 175–95.

Kendall, Frances. 2003. "How to Be an Ally if You Are a Person with Privilege" Accessed July 13, 2019. http://www.scn.org/friends/ally.html.

Khoury, Lamya, Yilang L. Tang, Bekh Bradley, Joe F. Cubells, and Kerry J. Ressler. 2010. "Substance Use, Childhood Traumatic Experience, and Posttraumatic Stress Disorder in an Urban Civilian Population." *Depression and Anxiety* 27 (12): 1077–86.

Kilbourne, Jean. 1999. *Can't Buy My Love: How Advertising Changes the Way We Think and Feel*. New York: Simon and Schuster.

Kimmel, Michael. 2000. *The Gendered Society*. New York: Oxford University Press.

Kimmerer, Robin Wall. 2013. *Braiding Sweetgrass: Indigenous Wisdom, Scientific Knowledge, and the Teachings of Plants*. Minneapolis: Milkweed.

Kirby, Andy. 2004. "Domestic Protest: The Eco-Village Movement as a Space of Resistance." *Bad Subjects*. Accessed December 19, 2014. http://bad.eserver.org/issues/2004/65/kirby.html.

Klein, Naomi. 2013. "Naomi Klein: Green Groups May be More Damaging than Climate Change Deniers. By Jason Mark." *Salon*. Accessed October 7, 2013. http://www.salon.com/2013/09/05/naomi_klein_big_green_groups_are_crippling_the_environmental_movement_partner/.

Klein, Naomi. 2014. *This Changes Everything: Capitalism vs. the Climate*. New York: Simon and Schuster.

Klerman, Gerald. L., and Myrna. M. Weissman. 1989. "Increasing Rates of Depression." *JAMA* 261 (15): 2229–35.

Kline, Neal A. 2006. "Revisiting Once upon a Time." *American Journal of Psychiatry* 163 (7): 1147–48.

Koont, Sinan. 2009. "The Urban Agriculture of Havana." *Monthly Review* 60 (8). Accessed March 10, 2013. http://monthlyreview.org/2009/01/01/the-urban-agriculture-of-havana.

Koont, Sinan. 2011. *Sustainable Urban Agriculture in Cuba*. Gainesville: University Press of Florida.

Kozeny, Geoph, ed. 1995. "Intentional Communities: Lifestyles Based on Ideals." In *Communities Directory: A Guide to Cooperative Living*. Rutledge, MO: Fellowship for Intentional Community, 18–24.

Krosnick, Jon, Allyson Holbrook, Laura Lowe, and Penny Visser. 2006. "The Origins and Consequences of Democratic Citizen's Policy Agendas: A Study of Popular Concern about Global Warming." *Climate Change* 77: 7–43.

Kuhn, Thomas. 2012. *The Structure of Scientific Revolutions: 50th Anniversary Edition*. 4th ed. Chicago: University of Chicago Press.

Labao, Linda. 2016. "The Sociology of Subnational Development: Conceptual and Empirical Foundations." In *The Sociology of Development Handbook*, edited by Gregory Hooks, 265–92. Oakland: University of California Press.

Ladd, Anthony E. 2014. "Environmental Disputes and Opportunity-Threat Impacts Surrounding Natural Gas Fracking in Louisiana." *Social Currents* 1 (3): 293–312.

Lal, R. 2004. "Soil Carbon Sequestration Impacts on Global Climate Change and Food Security." *Science* 304: 1623–27.

Lapidus, G. W. 1978. *Women in Soviet Society: Equality, Development, and Social Change*. Berkeley: University of California Press.

Larsen, Gary L. 2009. "An Inquiry Into the Theoretical Basis of Sustainability: Ten Propositions." In *Understanding the Social Dimensions of Sustainability*, edited by Dillard, Jesse, Veronica Dujon, and Mary C. King, 45–82. New York: Routledge.

Layard, Richard. 2005. *Happiness: Lessons from a New Science*. New York: Penguin Books.

Lazard. 2018. "Levelized Cost of Energy and Levelized Cost of Storage 2018." Accessed July 1, 2019. https://www.lazard.com/perspective/levelized-cost-of-energy-and-levelized-cost-of-storage-2018/.

Le Quéré, Corinne, Robbie M. Andrew, Pierre Friedlingstein, Stephen Sitch, Judith Hauck, Julia Pongratz, Penelope A. Pickers, et al. 2018. "Global Carbon Budget 2018." *Earth System Science Data* 10: 1–54. https://doi.org/10.5194/essd-10-2141-2018.

Lefebvre, Henri. 1996. "The Right to the City." In *Writings on Cities*, translated and edited by Eleonore Kofman and Elizabeth Lebas, 63–184. Blackwell Publishers Inc.: Malden Massachusetts.

Leibler, Jessica, Patricia A. Janulewicza, and Melissa J. Perry. 2017. "Prevalence of Serious Psychological Distress among Slaughterhouse Workers at a United States Beef Packing Plant." *Work* 57: 105–9.

Leibler, Jessica H., and Melissa J. Perry 2017. "Self-reported Occupational Injuries Among Industrial Beef Slaughterhouse Workers in the Midwestern United States." *Journal of Occupational and Environmental Hygiene* 14(1): 23–30. doi:10.1080/15459624.2016.1211283.

Leiserowitz, Anthony, Edward. Maibach, Connie Roser-Renouf, Geoff Feinberg, and Seth Rosenthal. 2015. "Climate Change in the American Mind: October, 2015." Yale University and George Mason University. New Haven: Yale Program on Climate Change Communication.

Leonard, Annie. 2010. *The Story of Stuff: The Impact of Overconsumption on the Planet, Our Communities, and Our Health—And How We Can Make It Better*. New York: Free Press.

Lerner, Steve. 2010. *Sacrifice Zones: The Front Lines of Toxic Chemical Exposure in the United States*. Cambridge: MIT Press.

Lewontin, R., and R. Levins. 2007. *Biology Under the Influence: Dialectical Essays on Ecology, Agriculture, and Health*. New York: Monthly Review Press.

Lidskog, Rolf, and Ingemar Elander. 2012. "Ecological Modernization in Practice? The Case of Sustainable Development in Sweden." *Journal of Environmental Policy & Planning* 14 (4): 411–27.

Litfin, Karen T. 2014. *Ecovillages: Lessons for Sustainable Community.* Malden, MA: Polity Press.

Logan, John R., and Harvey L. Molotch. 1987. *Urban Fortunes: The Political Economy of Place.* Berkeley: University of California Press.

Longo, Stefano B., Rebecca Clausen, and Brett Clark. 2015. *The Tragedy of the Commodity: Oceans, Fisheries, and Aquaculture.* New Brunswick, NJ: Rutgers University Press.

López, Victoria, and Sandra Park 2018. "ICE Detention Center Says It's Not Responsible for Staff's Sexual Abuse of Detainees" ACLU. Accessed October 7, 2020. https://www.aclu.org/blog/immigrants-rights/immigrants-rights-and-detention/ice-detention-center-says-its-not-responsible.

Lubitow, Amy, and Jennifer Allen. 2013. "Regulatory Barriers to Social Sustainability: Overcoming Environmental Health Policy Pitfalls." In *Social Sustainability: A Multilevel Approach to Social Inclusion*, edited by Jesse Dillard, Eileen Brennan, and Veronica Dujon, 79–102. New York: Routledge.

Luciak, Ilja A. 2007. *Gender and Democracy in Cuba.* Gainesville: University Press of Florida.

Lydeard, Charles, Robert H. Cowie, Winston F. Ponder, Arthur E. Bogan, Philippe Bouchet, Stephanie A. Clark, Kevin S. Cummings, et al. 2004. "The Global Decline of Nonmarine Mollusks." *BioScience* 54 (4): 321–30.

Maal-Bared, Rasha. 2005. "Comparing Environmental Issues in Cuba Before and After the Special Period: Balancing Sustainable Development and Survival." *Environment International* 32 (3): 349–58.

Magdoff Fred, John B. Foster, and Frederick H. Buttel. 2000. *Hungry for Profit: The Agribusiness Threat to Farmers, Food, and the Environment.* New York: Monthly Review Press.

Magis, Kristen and Craig Shinn. 2009. "Emergent Principles of Social Sustainability." In *Understanding the Social Dimensions of Sustainability*, edited by Jesse Dillard, Veronica Dujon, and Mary C. King, 15–44. New York: Routledge.

Malin, Stephanie. A. 2015. *The Price of Nuclear Power: Uranium Communities and Environmental Justice.* New Brunswick, NJ: Rutgers University Press.

Mark, Jason. 2013. "Naomi Klein: Green Groups May be More Damaging than Climate Change Deniers." *Salon.* Accessed October 7, 2013. http://www.salon.com/2013/09/05/naomi_klein_big_green_groups_are_crippling _the_environmental_movement_partner/.

Marshall, Bob. 2014. "A Football Field-Sized Area of Land Is Being Washed Away Every Hour, and Lawsuits Are Being Filed to Hold Oil and Gas Companies Responsible for the Destruction." *Scientific American.* Accessed September 7, 2017. https://www.scientificamerican.com/article/losing-ground-southeast-louisiana-is-disappearing-quickly/.

Maystadt, Jean-François, and Olivier Ecker. 2014. "Extreme Weather and Civil War: Does Drought Fuel Conflict in Somalia through Livestock Price Shocks?" *American Journal of Agricultural Economics* 96 (5): 1157–82.

Mazmanian, Daniel A., and Michael E. Kraft. 2009. "Introduction." In *Toward Sustainable Communities: Transition and Transformation in Environmental Policy*, edited by Daniel A. Mazmanian and Michael E. Kraft, 1–2. Cambridge, MA: MIT Press.

McAnany, Patricia, and Norman Yoffee. 2010. "Why We Question Collapse and Study Human Resilience, Ecological Vulnerability, and the Aftermath of Empire." In

Questioning Collapse: Human Resilience, Ecological Vulnerability, and the Aftermath of Empire, edited by Patricia McAnany and Norman Yoffee. New York: Cambridge University Press.

McClintock, Nathan. 2014. "Radical, Reformist, and Garden-Variety Neoliberal: Coming to Terms with Urban Agriculture's Contradictions." *Local Environment: The International Journal of Justice and Sustainability* 19 (2): 147–71.

McEvers, Kelly. 2015. "Utah Reduced Chronic Homelessness by 91 Percent; Here's How." *NPR*. Accessed July 3, 2019. https://www.npr.org/2015/12/10/459100751/utah-reduced-chronic-homelessness-by-91-percent-heres-how.

McGee, Julius Alexander. 2015. "Does Certified Organic Farming Reduce Greenhouse Gas Emissions from Agricultural Production?" *Agriculture and Human Values* 32 (2): 255–63.

McGee, Julius Alexander, and Camila Alvarez. 2016. "Sustaining without Changing: The Metabolic Rift of Certified Organic Farming." *Sustainability* 8: 115.

McGee, Julius, Christina Ergas, and Matt Clement. 2018. "Racing to Reduce Emissions: Assessing the Relationship between Race and Carbon Dioxide Emissions from On-Road Travel." *Sociology of Development*.

McMichael, Philip. 2010a. *Contesting Development: Critical Struggles for Social Change.* New York: Routledge.

McMichael, Philip. 2010b. "The World Food Crisis in Historical Perspective." In *Agriculture and Food in Crisis: Conflict, Resistance, and Renewal*, edited by Fred Magdoff and Brian Tokar, 51–68. New York: Monthly Review Press.

McNeill, John R. 2000. *Something New under The Sun: An Environmental History of the Twentieth-Century World.* New York: W. W. Norton.

McPherson, Guy. 2018. "On Imminent Human Extinction: Interviewed by Rajani Kanth." Nature Bats Last. Accessed July 1, 2019. https://guymcpherson.com/2018/10/on-imminent-human-extinction-interviewed-by-rajani-kanth/.

McPherson, Miller, Lynn Smith-Lovin, and Matthew E. Brashears. 2006. "Social Isolation in America: Changes in Core Discussion Networks over Two Decades." *American Sociological Review* 71 (3): 353–75.

Melosi, Martin V. 2010. "Humans, Cities, and Nature: How Do Cities Fit in the Material World?" *Journal of Urban History* 36 (1): 3–21.

Melucci, Alberto. 1995. "The Process of Collective Identity." In *Social Movements and Culture: Protests and Contention*, 4th ed., edited by Hank Johnston and Bert Klandermans, 41–63. Minneapolis: University of Minnesota Press.

Merchant, Carolyn. 1980. *The Death of Nature: Women, Ecology, and the Scientific Revolution.* New York: HarperCollins Publishers.

Merica, Dan. 2017. "What Trump's Climate Change Order Accomplishes—and What It Doesn't." *CNN*. Accessed April 18, 2017. http://www.cnn.com/2017/03/28/politics/donald-trump-climate-change-executive-order/.

Merriam-Webster Dictionary. 2019. "Regeneration." Accessed July 13, 2019. https://www.merriam-webster.com/dictionary/regeneration.

Mies, Maria. 1998. *Patriarchy and Accumulation on a World Scale: Women in the International Division of Labor.* New York: Zed Books.

Mies, Maria, and Vandana Shiva. 2014. *Ecofeminism*. New York: Zed Books.

Milan, Andrea, Robert Oakes, and Jillian Campbell. 2016. "Tuvalu: Climate Change and Migration: Relationships Between Household Vulnerability, Human Mobility and

Climate Change." United Nations University Institute for Environment and Human Security. Accessed October 7, 2020. https://environmentalmigration.iom.int/tuvalu-climate-change-and-migration-relationships-between-household-vulnerability-human-mobility-and

Milanovic, Branko. 2005. *Worlds Apart: Measuring International and Global Inequality.* Princeton, NJ: Princeton University Press.

Mohai, Paul, David Pellow, and J. Timmons Roberts. 2009. "Environmental Justice." *Annual Review of Environment and Resources*, 34: 405–30.

Mol, Arthur P. J. 2001. *Globalization and Environmental Reform: The Ecological Modernization of the Global Economy.* Cambridge, MA: MIT Press.

Momsen, Janet H. 2010. *Gender and Development.* 2nd ed. London: Routledge.

Molotch, Harvey. 1976. "The City as Growth Machine: Toward a Political Economy of Place." *American Journal of Sociology* 82 (2): 309–32.

Molotch, Harvey, and John Logan. 1984. "Tensions in the Growth Machine: Overcoming Resistance to Value-Free Development." *Social Problems* 31 (5): 483–99.

Moore, Jason. 2000. "Environmental Crises and the Metabolic Rift in World-Historical Perspective." *Organization & Environment* 13 (2): 123–57.

Moore, Michael J., Carolyn W. Zhu, and Elizabeth C. Clipp. 2001. "Informal Costs of Dementia Care: Estimates from the National Longitudinal Caregiver Study." *The Journal of Gerontology: Series B* 56 (4): S219–S228.

Moore, Steven. 2007. *Alternative Routes to the Sustainable City: Austin, Curitiba, and Frankfurt.* United Kingdom: Lexington Books.

Morello-Frosch, Rachel, and Bill M. Jesdale. 2006. "Separate and Unequal: Residential Segregation and Estimated Cancer Risks Associated With Ambient Air Toxics in U.S. Metropolitan Areas." *Environmental Health Perspectives* 114 (3): 386–93.

Morello-Frosch, Rachel, and Russ Lopez. 2006. "The Riskscape and the Color Line: Examining the Role of Segregation in Environmental Health Disparities." *Environmental Resources*, 102: 181–96.

Morgan, Faith. 2006. The Power of Community: How Cuba Survived Peak Oil. Yellow Springs, OH: Community Solutions. http://www.communitysolution.org/mediaandeducation/films/powerofcommunity/#.

Mosley, Stephen. 2010. *The Environment in World History.* New York: Routledge.

Mumford, Lewis. 1961. *The City in History: Its Origins, Its Transformations, and Its Prospects.* New York: Harcourt, Brace and World.

NASA. 2017. "NASA, NOAA Data Show 2016 Warmest Year on Record Globally." Accessed April 13, 2017. https://www.nasa.gov/press-release/nasa-noaa-data-show-2016-warmest-year-on-record-globally.

NASA. 2019. "Global Climate Change: Vital Signs of the Planet." Accessed April 8, 2019. https://climate.nasa.gov/vital-signs/global-temperature/

NASA. 2021. "2020 Tied for Warmest Year on Record, NASA Analysis Shows." Accessed April 28, 2021 https://www.nasa.gov/press-release/2020-tied-for-warmest-year-on-record-nasa-analysis-shows

NASA Goddard Institute for Space Studies. 2017. "January 2017 Was Third-Warmest January on Record." Accessed April 18, 2017. https://climate.nasa.gov/news/2550/january-2017-was-third-warmest-january-on-record/.

National Institute of Mental Health. 2017. "Mental Illness." Statistics. Accessed October 25, 2018. https://www.nimh.nih.gov/health/statistics/mental-illness.shtml.

National Institute on Drug Abuse. 2015. "Nationwide Trends." Drug Facts. Accessed October 25, 2018. https://www.drugabuse.gov/publications/drugfacts/nationwide-trends.

National Oceanic Atmospheric Association. 2017. "What Is Coral Bleaching?" Accessed September 7, 2017. https://oceanservice.noaa.gov/facts/coral_bleach.html.

National Oceanic Atmospheric Association. 2019. "2018 Was 4th Hottest Year on Record for the Globe: The U.S. Experienced 14 Billion-Dollar Weather and Climate Disasters." Accessed April 8, 2019. https://www.noaa.gov/news/2018-was-4th-hottest-year-on-record-for-globe.

Natural Resources Defense Council (NRDC). 2008. "The Cost of Climate Change: What We'll Pay if Global Warming Continues Unchecked." www.nrdc.org/policy.

Nature Climate Change. 2018. "Focus on Climate Change and Mental Health." *Nature Climate Change* 8: 259. Editorial. https://doi.org/10.1038/s41558-018-0128-7.

Neumann, Roderick P. 1998. *Imposing Wilderness Struggles over Livelihood and Nature Preservation in Africa*. Berkeley: University of California Press.

New Economics Foundation. 2016. "Happy Planet Index." Accessed September 21, 2017. http://www.happyplanetindex.org/data/.

Norgaard, Kari Marie. 2009. "Cognitive and Behavioral Challenges in Responding to Climate Change." The World Bank Development Economics World Development Report Team May 2009. *Policy Research Working Paper 4940*. Washington, DC: World Bank.

Norgaard, Kari Marie. 2011. *Living in Denial: Climate Change, Emotions, and Everyday Life*. Cambridge, MA: MIT Press.

Norgaard, Kari Marie. 2015. "Normalizing the Unthinkable: Climate Denial in Everyday Life." In *Twenty Lessons in Environmental Sociology*, edited by Kenneth Gould and Tammy Lewis, 246–59. New York: Oxford University Press.

Norgaard, Kari M., Ron Reed, and Caroline Van Horn. 2011. "A Continuing Legacy: Institutional Racism, Hunger, and Nutritional Justice on the Klamath." In *Cultivating Food Justice: Race, Class, and Sustainability*, edited by Alison H. Alkon and Julian Agyeman. Cambridge, MA: MIT Press.

Norgaard, Kari, and Richard York. 2005. "Gender Equality and State Environmentalism." *Gender and Society* 19 (4): 506–22.

Norton, Michael I., and Dan Ariely. 2011. "Building a Better America: One Wealth Quintile at a Time." *Perspectives on Psychological Science* 6 (1): 9–12.

Nowak, Martin. 2006. "Five Rules for the Evolution of Cooperation." *Science* 314 (5805): 1560–63.

Nowrasteh, Alex. 2020. "8 People Died in Immigration Detention in 2019, 193 Since 2004." CATO Institute. Accessed October 7, 2020. https://www.cato.org/blog/8-people-died-immigration-detention-2019-193-2004.

Obradovich, Nick, Robyn Migliorini, Martin P. Paulus, and Iyad Rahwan. 2018. "Empirical Evidence of Mental Health Risks Posed by Climate Change." *PNAS* 115 (43): 10953–58. https://doi.org/10.1073/pnas.1801528115.

Oficina Nacional de Estadisticas (ONEI) República de Cuba. 2011. "3.7 – Población Residente y Densidad de Población por Provincias, Según Zonas Urbana y Rural." Accessed April 11, 2012. http://www.onei.gob.cu.

Oliver, J. Eric. 2003. "Mental Life and the Metropolis in Suburban America: The Psychological Correlates of Metropolitan Place Characteristics." *Urban Affairs Review* 39 (2): 228–53.

"The 100 Most Important Americans of the 20th Century: Special Issue." 1990. *Life Magazine*. Accessed October 5, 2017. http://www.originallifemagazines.com/LIFE-Magazine-Fall-1990-P3030.aspx.

Opotow, Susan. 1990. "Moral Exclusion and Injustice: An Introduction." *Journal of Social Issues* 46 (1): 1–20.

Opotow, Susan. 1994. "Predicting Protection: Scope of Justice and the Natural World." *Journal of Social Issues* 50 (3): 49–63.

Owen, David. 2009. *Green Metropolis: Why Living Smaller, Living Closer, and Driving Less Are the Keys to Sustainability.* New York: Penguin.

Øyen, Else (ed.). 1990. "The Imperfection of Comparison." In *Comparative Methodology: Theory and Practice in International Social Research*, 1–18. London: Sage.

Parr, Adrian. 2009. *Hijacking Sustainability.* Cambridge, MA: MIT Press.

Pages, Raisa. 2008. "The Status of Cuban Women: From Economically Dependent to Independent." In *A Contemporary Cuba Reader: Reinventing the Revolution*, edited by Philip Brenner, Marguerite Rose Jimenez, John. M. Kirk, and William M. LeoGrande, 311–15. New York: Rowman and Littlefield.

Pelejero, Carles, Eva Calvo, and Ove Hoegh-Guldberg. 2010. "Paleo-Perspectives on Ocean Acidification." *Trends in Ecological Evolution* 25 (6): 332–44.

Pellow, David Naguib. 2002. *Garbage Wars: The Struggle for Environmental Justice in Chicago.* Cambridge, MA: MIT Press.

Pellow, David Naguib. 2007. *Resisting Global Toxics: Transnational Movements for Environmental Justice.* Cambridge, MA: MIT Press.

Pellow, David Naguib. 2014. *Total Liberation: The Power and Promise of Animal Rights and the Radical Earth Movement.* Minneapolis: University of Minnesota Press.

Pellow, David Naguib. 2018. *What is Critical Environmental Justice?* Medford, MA: Polity Press.

Penna, Anthony N. 2010. *The Human Footprint: A Global Environmental History.* Malden, MA: Wiley-Blackwell.

Pew Research Center. 2015. "Modern Immigration Wave Brings 59 Million to U.S., Driving Population Growth and Change through 2065." Hispanic Trends. Accessed October 7, 2020. https://www.pewresearch.org/hispanic/2015/09/28/modern-immigration-wave-brings-59-million-to-u-s-driving-population-growth-and-change-through-2065/

Pfeiffer, Dale. 2006. *Eating Fossil Fuels: Oil, Food, and the Coming Crisis in Agriculture.* Gabriola Island, BC, Canada: New Society.

Pierce, Sarah. 2015. "Forced Labor in US Detention Centers." Human Trafficking Search. Accessed October 7, 2020. https://humantraffickingsearch.org/forced-labor-in-u-s-detention-centers/.

Pieters, Rik. 2013. "Bidirectional Dynamics of Materialism and Loneliness: Not Just a Vicious Cycle." *Journal of Consumer Research* 40 (4): 615–31.

Pimm, Stuart L., Clinton N. Jenkins, Robin Abell, Thomas M. Brooks, John L. Gittleman, Lucas N. Joppa, Peter. H. Raven, Callum. M. Roberts, and Joseph. O. Sexton. 2014. "The Biodiversity of Species and Their Rates of Extinction, Distribution, and Protection." *Science* 344 (6187): 1246752.

Pimm, Stuart L., and Peter Raven. 2000. "Biodiversity: Extinction by Numbers." *Nature* 403: 843–45.

Pimm, Stuart L., Gareth J. Russell, John L. Gittleman, and Thomas M. Brooks. 1995. "The Future of Biodiversity." *Science* 269 (5222): 347–50.

Pinquart, Martin, and Silvia Sorensen. 2003. "Differences between Caregivers and Noncaregivers in Psychological Health and Physical Health: A Meta-Analysis." *Psychology and Aging* 18 (2): 250–67.

Plumwood, Val. 2002. *Environmental Culture: The Ecological Crisis of Reason.* New York: Routledge.

Polanyi, Karl. 2001. *The Great Transformation: The Political and Economic Origins of Our Lives.* Boston, MA: Beacon Press.

Polimeni, John M., Kozo Mayumi, Mario Giampietro, and Blake Alcott. 2008. *The Jevons Paradox and the Myth of Resource Efficiency Improvements.* London: Earthscan.

Pollan, Michael. 2006. *The Omnivore's Dilemma: A Natural History of Four Meals.* New York: Penguin.

Ponting, Clive. 2007. *A New Green History of the World: The Environment and the Collapse of Great Civilizations.* New York: Penguin.

Portney, Kent E. 2003. *Taking Sustainable Cities Seriously: Economic Development, the Environment, and Quality of Life in American Cities.* Cambridge, MA: MIT Press.

Premat, Adriana. 2005. "Moving Between the Plan and the Ground: Shifting Perspectives on Urban Agriculture in Havana, Cuba." In *Agropolis: The Social, Political and Environmental Dimensions of Urban Agriculture*, edited by Luc J. A. Mougeot, 153–86. Sterling, VA: Earthscan and the International Development Research Centre.

Presser, Lois. 2008. *Been a Heavy Life: Stories of Violent Men.* Urbana and Chicago: University of Illinois Press.

Presser, Lois. 2013. *Why We Harm.* New Brunswick, NJ: Rutgers University Press.

Presser, Lois, and Sveinung Sandberg. 2015. "Research Strategies for Narrative Criminology." In *Advances in Criminological Theory: The Value of Qualitative Research for Advancing Criminological Theory*, edited by Jody Miller and Wilson R. Palacios, 85–99. Piscataway, NJ: Transaction.

Prieto, Rodrigo Espina, and Pablo Rodríguez Ruiz. 2010. "Race and Inequality in Cuba Today." *Socialism and Democracy* 24 (1): 161–77.

Public Citizen. 2017. "Storm of Silence: Media Coverage of Climate Change and Hurricane Harvey." Accessed October 30, 2017. https://www.citizen.org/system/files/case_documents/public_citizen_storm_of_silence_harvey_climate_coverage_1.pdf.

Pulido, Laura. 2017. "Geographies of Race and Ethnicity II: Environmental Racism, Racial Capitalism and State-Sanctioned Violence." *Progress in Human Geography* 41 (4): 524–33.

Putnam, Robert. D. 1995. "Bowling Alone: America's Declining Social Capital." *Journal of Democracy* 6 (1): 65–78.

Putnam, Robert. D. 2000. *Bowling Alone: The Collapse and Revival of American Community.* New York: Simon and Schuster.

Quastel, Noah. 2009. "Political Ecologies of Gentrification." *Urban Geography* 30 (7): 694–725.

Raby, Diana. 2009. "Why Cuba Still Matters." *Monthly Review* 60 (8). Accessed March 10, 2013. http://monthlyreview.org/2009/01/01/why-cuba-still-matters.

Radley, David, Sara Collins, and Susan Hayes. 2019. "2019 Scorecard on State Health System Performance." The Commonwealth Fund. Accessed July 13, 2019. www.datacenter.commonwealthfund.org.

Rafferty, John. 2020. "Just How Old Is Homo Sapiens?" *Britannica*. Accessed October 27, 2020. https://www.britannica.com/story/just-how-old-is-homo-sapiens

Rees, William, and Mathis Wackernagel. 1996. "Urban Ecological Footprints: Why Cities Cannot Be Sustainable—And Why They Are a Key to Sustainability." *Environment Impact Assessment Review* 16: 223–48.

Régniera, Claire, Guillaume Achaz, Amaury Lambertd, Robert H. Cowieg, Philippe Boucheta, and Benoît Fontaineh. 2015. "Mass Extinction in Poorly Known Taxa." *PNAS* 112 (25): 7761–66.

Reporters Without Borders. 2012. "Press Freedom Index 2011–2012." Accessed September 21, 2013. https://rsf.org/en/world-press-freedom-index-20112012.

Reuters. 2012. "Cuba Reports Food Output Up 8.7 Percent in 2011." *Reuters.* Accessed October 8, 2013. http://www.reuters.com/article/food-cuba-idUSN1E81F01520120216.

Richards, Lisa, Nigel Brew, and Lizzie Smith. 2020. "2019–20 Australian Bushfires—Frequently Asked Questions: A Quick Guide" Research Papers 2019–20. Parliament of Australia. Accessed October 7, 2020. https://www.aph.gov.au/About_Parliament/Parliamentary_Departments/Parliamentary_Library/pubs/rp/rp1920/Quick_Guides/AustralianBushfires.

Richardson, John H. 2018. "When the End of Human Civilization Is Your Day Job: Among Many Climate Scientists, Gloom Has Set in. Things Are Worse than We Think, but They Can't Really Talk about It." *Esquire.* Accessed July 1, 2019. https://www.esquire.com/news-politics/a36228/ballad-of-the-sad-climatologists-0815/.

Rick, Scott I., Beatriz Pereira, Katherine A. Burson. 2014. "The Benefits of Retail Therapy: Making Purchase Decisions Reduces Residual Sadness." *Journal of Consumer Psychology* 24 (3): 373–80.

Rockström, Johan, Owen Gaffney, Joeri Rogelj, Malte Meinshausen, Nebojsa Nakicenovic, and Hans Joachim Schellnhuber. 2017. "A Roadmap for Rapid Decarbonization: Emissions Inevitably Approach Zero with a 'Carbon Law.'" *Science* 355 (6331): 1269–71.

Rogers, Heather. 2010a. *Green Gone Wrong: Dispatches from the Front Lines of Eco-Capitalism.* New York: Verso.

Rogers, Heather. 2010b. *Green Gone Wrong: How Our Economy Is Undermining the Environmental Revolution.* New York: Scribner.

Romm, Joseph. 2011. "Desertification: The Next Dust Bowl." *Nature* 478: 450–51.

Rosenberg, Marshall. 2005. *Nonviolent Communication: A Language of Life.* Encinitas, CA: PuddleDancer Press.

Rosin, Hanna. 2009. "Did Christianity Cause the Crash?" *The Atlantic.* Accessed October 8, 2017. https://www.theatlantic.com/magazine/archive/2009/12/did-christianity-cause-the-crash/307764/.

Rosset, Peter. M. 2000. "Cuba: A Successful Case Study of Sustainable Agriculture." In *Hungry for Profit: The Agribusiness Threat to Farmers, Food, and the Environment,* edited by Fred Magdoff, John Bellamy Foster, and Frederick H. Buttel, 203–13. New York: Monthly Review Press.

Rosset, Peter, Braulio Machín-Sosa, Adilén M. Roque-Jaime, and Dana R. Avila-Lozano. 2011. "The Campesino-to-Campesino Agroecology Movement of ANAP in Cuba." *Journal of Peasant Studies* 38: 161–91.

Rostow, Walt Whitman. 1959. "The Stages of Economic Growth." *The Economic History Review* 12 (1): 1–16.

Rudel, Thomas. 2009. "How Do People Transform Landscapes?: A Sociological Perspective on Suburban Sprawl and Tropical Deforestation." *American Journal of Sociology*. 115 (1):129–54.

Rumbaut, Luis, and Rubén G. Rumbaut. 2009. "Survivor: Cuba: The Cuban Revolution at 50." *Latin American Perspectives* 36 (1): 84–98.

Rushkoff, Douglas. 2018. "How Tech's Richest Plan to Save Themselves After the Apocalypse." *The Guardian*. Accessed September 26, 2018. https://www.theguardian.com/technology/2018/jul/23/tech-industry-wealth-futurism-transhumanism-singularity.

Sadler, Berry. 1999. "A Framework for Environmental Sustainability Assessment and Assurance." In *Handbook of Environmental Impact Assessment, Volume 1: Environmental Impact Assessment: Process, Methods, and Potential*, edited by Judith Petts, 12–32. Malden, MA: Blackwell Science.

Salleh, Ariel. 2010. "From Metabolic Rift to Metabolic Value: Reflections on Environmental Sociology and the Alternative Globalization Movement." *Organization & Environment* 23 (2): 205–19.

Satterthwaite, David. 2009. "The Implications of Population Growth and Urbanization for Climate Change." *Environment and Urbanization* 21 (2): 545–67.

Satterthwaite, David. 2010. "The Contribution of Cities to Global Warming and Their Potential Contributions to Solutions." *Environment and Urbanization Asia* 1 (1): 1–12.

Sbicca, Joshua. 2012. "Growing Food Justice by Planting an Anti-Oppression Foundation: Opportunities and Obstacles for a Budding Social Movement." *Agriculture and Human Values* 29 (4): 455–66.

Sbicca, Joshua. 2014. "The Need to Feed: Urban Metabolic Struggles of Actually Existing Radical Projects." *Critical Sociology* 40 (6): 817–34.

Schaef, Anne W. 1987. *When Society Becomes an Addict*. San Francisco: Harper and Row.

Scheer, Roddy, and Doug Moss. 2017. "Sustainability: Use It and Lose It: The Outsize Effect of U.S. Consumption on the Environment." *Scientific American*. Accessed October 15, 2017. https://www.scientificamerican.com/article/american-consumption-habits/#.

Schehr, Robert C. 1997. *Dynamic Utopia: Establishing Intentional Communities as a New Social Movement*. Westport, CT: Bergin and Garvey.

Schnaiberg, Allan. 1980. *The Environment: From Surplus to Scarcity*. New York: Oxford University Press.

Schor, Juliet. 1998. *The Overspent American: Upscaling, Downshifting, and the New Consumer*. New York: Basic Books.

Schulz, Richard, and Scott Beach. 1999. "Caregiving as a Risk Factor for Mortality: The Caregiver Health Effects Study." *American Medical Association* 282 (23): 2215–19.

Scott, Shaunna, Stephanie McSpirit, Patrick Breheny, and Britteny M. Howell. 2012. "The Long-Term Effects of a Coal Waste Disaster on Social Trust in Appalachian Kentucky." *Organization & Environment* 25 (4): 402–18.

Sekerci, Yadigar, and Sergei Petrovskii. 2015. "Mathematical Modelling of Plankton-Oxygen Dynamics under the Climate Change." *Bulletin of Mathematical Biology* 77 (12): 2325–53.

Seyfang, Gill. 2010. "Community Action for Sustainable Housing: Building a Low-Carbon Future." *Energy Policy* 38: 7624–33.

Seyfang, Gill, and Adrian Smith. 2007. "Grassroots Innovations for Sustainable Development: Towards a New Research and Policy Agenda." *Environmental Politics* 16 (4): 584–603.

Shayne, Julie D., 2004. *The Revolution Question: Feminism in El Salvador, Chile, and Cuba.* New Brunswick, NJ: Rutgers University Press.

Sherwood, Steven C., and Matthew Huber. 2010. "An Adaptability Limit to Climate Change Due to Heat Stress." *Proceedings of the National Academy of Sciences of the United States of America* 107 (21): 9552–55.

Shiva, Vandana. 2005. *Earth Democracy: Justice, Sustainability, and Peace.* New York: South End Press.

Shiva, Vandana. 2016. *Who Really Feeds the World?: The Failures of Agribusiness and the Promise of Agroecology.* Berkeley, CA: North Atlantic Books.

SIPRI. 2017. "World Military Spending: Increases in the USA and Europe, Decreases in Oil-Exporting Countries." Accessed October 11, 2017. https://www.sipri.org/media/press-release/2017/world-military-spending-increases-usa-and-europe.

Sirna, Tony. 2016. "Cutting Our Carbon Footprint." Dancing Rabbit Ecovillage. Accessed November 11, 2017. https://www.dancingrabbit.org/about-dancing-rabbit-ecovillage/eco-living/cutting-our-carbon-footprint/.

Slater, Philip. 1970. *The Pursuit of Loneliness: America's Discontent and the Search for a New Democratic Ideal.* Boston: Beacon Press.

Smil, Vaclav. 1994. "How Many People Can the Earth Feed?" *Population and Development Review.* 20 (2):255–92.

Smith, Barry E. 2002a. "Structure: Nitrogenase Reveals Its Inner Secrets." *Science* 297: 1654–55.

Smith, Dorothy E. 1987. *The Everyday World as Problematic: A Feminist Sociology.* Boston: Northeastern University Press.

Smith, H. Jesse. 2015. "Another Cause of Climate Change Is Developing." *Science* 347 (6221): 516.

Smith, Lois M., and Alfred Padula. 1996. *Sex and Revolution: Women in Socialist Cuba.* New York: Oxford University Press.

Smith, William L. 2002b. "Intentional Communities 1990–2000: A Portrait." *Michigan Sociological Review* 16: 107–31.

Spaargaren, Gert, Peter Oosterveer, and Anne Loeber. 2012. *Food Practices in Transition: Changing Food Consumption, Retail and Production in the Age of Reflexive Modernity.* New York: Routledge.

Spratt, David, and Ian Dunlop. 2019. *Existential Climate-Related Security Risk: A Scenario Approach.* Policy Paper. Melbourne, Australia: Breakthrough National Centre for Climate Restoration.

Steffen, Will, Wendy Broadgate, Lisa Deutsch, Owen Gaffney, and Cornelia Ludwig. 2015. "The Trajectory of the Anthropocene: The Great Acceleration." *The Anthropocene Review* 2 (1): 81–98.

Steffen, Will, Paul J. Crutzen, and John R. McNeill. 2007. "The Anthropocene: Are Humans Now Overwhelming the Great Forces of Nature?" *Ambio* 36 (8): 614–21.

Stern, David I., and Robert K. Kaufmann. 2014. "Anthropogenic and Natural Causes of Climate Change." *Climatic Change* 122: 257–69.

Sternbach, Nancy, Marysa Navarro-Aranguren, Patricia Chuchryk, and Sonia Alvarez. 1992. "Feminisms in Latin America: From Bogota to San Bernardo." In *The Making of Social Movements in Latin America: Identity Strategy, and Democracy*, edited by Arturo Escobar and Sonia E. Alvarez, 207–39. Boulder, CO: Westview.

Stewart, John M. 1992. *The Soviet Environment: Problems, Policies, and Politics.* Cambridge and New York: Cambridge University Press.

Stiglitz, Joseph. 2003. *Globalization and Its Discontents.* New York: W. W. Norton.

Stricker, Pamela. 2007. *Toward a Culture of Nature: Environmental Policy and Sustainable Development in Cuba.* Latham, MD: Lexington Books.

Stricker, Pamela. 2010. "Bringing Social Justice Back in: Cuba Revitalizes Sustainable Development." *Local Environment* 15 (2): 185–97.

Stromberg, Joseph. 2013. "What Is the Anthropocene and Are We in It?: Efforts to Label the Human Epoch Have Ignited a Scientific Debate Between Geologists and Environmentalists." *Smithsonian Magazine.* Accessed June 27, 2016. http://www.smithsonianmag.com/science-nature/what-is-the-anthropocene-and-are-we-in-it-164801414/#kKxXzqxJOTOPwO5r.99.

Stuart, Graeme. 2014. "What Is Community Capacity Building?" *Sustaining Community: Families, Communities, the Environment.* Accessed July 13, 2019. https://sustainingcommunity.wordpress.com/2014/03/10/ccb/.

Studdert, David. 2005. *Conceptualising Community: Beyond the State and Individual.* New York: Palgrave Macmillan.

Sturm, Roland, and Deborah A. Cohen. 2004. "Suburban Sprawl and Physical and Mental Health." *Public Health* 118 (7): 488–96.

Subramanian, Subu, and Ichiro Kawachi. 2004. "Income Inequality and Health: What Have We Learned So Far?" *Epidemiologic Reviews* 26 (1): 78–91. https://doi.org/10.1093/epirev/mxh003.

Supran, Geoffrey, and Naomi Oreskes. 2017. "Assessing ExxonMobil's Climate Change Communications (1977–2014)." *Environmental Research Letters* 12 (8). https://doi.org/10.1088/1748-9326/aa815f.

Swidler, Ann. 1986. "Culture in Action: Symbols and Strategies." *American Sociological Review* 51, no. 2 (April): 273–86.

Thatcher, Margaret. 1987. Interview for Woman's Own ("no such thing as society"). The Margaret Thatcher Foundation. https://www.margaretthatcher.org/document/106689

Thompson, Craig J., and Juliet Schor. 2014. "Cooperative Networks, Participatory Markets, and Rhizomatic Resistance: Situating Plentitude Within Contemporary Political Economy Debates." In *Sustainable Lifestyles and the Quest for Plentitude: Case Studies of the New Economy*, edited by Juliet Schor and Craig J. Thompson, 233–50. New Haven, CT: Yale University Press.

Tickamyer, Ann R., and Anouk Patel-Campillo. 2016. "Sociological Perspectives on Uneven Development: The Making of Regions." In *The Sociology of Development Handbook*, edited by Gregory Hooks, 293–310. Oakland: University of California Press.

Tierney, Kathleen. 2015. "Resilience and the Neoliberal Project: Discourses, Critiques, Practices—And Katrina." *American Behavioral Scientist* 59 (10): 1327–42.

Tolentino, Jia. 2017. "The Gig Economy Celebrates Working Yourself to Death." *The New Yorker.* Accessed October 15, 2017. https://www.newyorker.com/culture/jia-tolentino/the-gig-economy-celebrates-working-yourself-to-death.

Treisman, Rachel. 2020. "Whistleblower Alleges 'Medical Neglect,' Questionable Hysterectomies of ICE Detainees." *NPR.* Accessed October 7, 2020. https://www.npr.org/2020/09/16/913398383/whistleblower-alleges-medical-neglect-questionable-hysterectomies-of-ice-detaine.

Trucost. 2013. "Natural Capital at Risk: The Top 100 Externalities of Business." TRUCOST PLC. www.trucost.com

Turner, Graham. 2014. *Is Global Collapse Imminent?* MSSI Research Paper No. 4. Melbourne, Australia: Melbourne Sustainable Society Institute, University of Melbourne.

Twenge, Jean M. 2015. "Time Period and Birth Cohort Differences in Depressive Symptoms in the U.S., 1982–2013." *Social Indicators Research* 121 (2): 437–54.

Twenge, Jean M., Brittany Gentile, C. Nathan DeWall, Debbie Ma, Katharine Lacefield, and David R. Schurtz. 2010. "Birth Cohort Increases in Psychopathology Among Young Americans, 1938–2007: A Cross-Temporal Meta-Analysis of the MMPI." *Clinical Psychology Review* 30: 145–54.

United Nations. 1987. "Our Common Future." *Report of the World Commission on Environment and Development (General Assembly Resolution 42/187)*. Accessed May 24, 2008. http://www.un.org/documents/ga/res/42/ares42-187.htm.

United Nations. 2005. "Ecosystems and Human Well-Being: Our Human Planet." *Millennium Ecosystem Assessment*. Accessed December 19, 2017. https://www.millenniumassessment.org/en/index.html.

United Nations. 2008. "World Urbanization Prospects: The 2007 Revision Executive Summary." Economic and Social Affairs. Accessed December 16, 2009. https://www.un.org/en/development/desa/population/events/pdf/expert/13/Heilig.pdf.

United Nations. 2012. *Population Division: World Urbanization Prospects, the 2011 Revision: Highlights*. Department of Economic and Social Affairs. New York. Accessed January 19, 2013. https://www.un.org/en/development/desa/population/publications/pdf/urbanization/WUP2011_Report.pdf.

United Nations. 2018. "Summary: Human Development Indices and Indicators." *2018 Statistical Update*. Accessed February 22, 2019. http://hdr.undp.org/en/2018-update.

United Nations Conference on Environment and Development. 1992a. *Agenda 21, The Rio Declaration on Environment and Development, and the Statement of Principles for the Sustainable Management of Forests*. Accessed November 27, 2011. http://www.un.org/esa/dsd/agenda21/.

United Nations Conference on Environment and Development. 1992b. *Earth Summit Agenda 21: The Statement of Principles*. Accessed September 26, 2013. http://sustainabledevelopment.un.org/content/documents/Agenda21.pdf.

United Nations Department of Economic and Social Affairs. *Population Division. 2010. World Urbanization Prospects, the 2009 Revision: Highlights*. New York. Accessed March 10, 2013. http://esa.un.org/unpd/wup/doc_highlights.htm.

United Nations Department of Economic and Social Affairs. 2012. Department of Economic and Social Affairs. *Population Division: World Urbanization Prospects, the 2011 Revision: Highlights*. New York. Accessed January 19, 2013. http://esa.un.org/unpd/wup/Documentation/highlights.htm

United Nations Development Programme. 1996. *Urban Agriculture: Food, Jobs, and Sustainable Cities*. New York: United Nations Development Programme.

United Nations Development Programme. 2007/2008. "Fighting Climate Change." *Human Development Report*. Accessed July 12, 2010. hdr.undp.org/en/media/HDR_20072008_EN_Complete.pdf.

United Nations Development Programme. 2008. *Capacity Assessment Methodology: User's Guide*. Capacity Development Group Bureau for Development Policy. Accessed July 13, 2019. www.capacity.undp.org.

United Nations Development Programme. 2010. *Regional Human Development Report for Latin America and the Caribbean 2010. Acting on the Future: Breaking the*

Intergenerational Transmission of Inequality. Accessed September 27, 2013. http://hdr.undp.org/en/reports/regional/featuredregionalreport/idhalc_ en_2010.pdf.

United Nations Development Programme. 2013. *Human Development Report 2013.* Accessed September 21, 2013. http://hdr.undp.org/en/ reports/global/hdr2013/.

United Nations Development Programme. 2016. *Human Development Report 2016: Human Development for Everyone.* Accessed October 12, 2017. hdr.undp.org/sites/default/files/2016_human_development_report.pdf.

United Nations Environment Programme. 2007. "Sudan Post-Conflict Environmental Assessment." Accessed December 19, 2017. https://postconflict.unep.ch/publications/UNEP_Sudan.pdf.

United Nations Environment Programme. 2012. "One Planet, How Many People? A Review of Earth's Carrying Capacity." UNEP Global Environmental Alert Service (GEAS): Taking the Pulse of the Planet; Connecting Science with Policy. Accessed April 28, 2021 https://na.unep.net/geas/getUNEPPageWithArticleIDScript.php?article_id=88.

United Nations Environment Programme. 2017. "30th Anniversary of Montreal Protocol and World Ozone Day 2017." Accessed September 9, 2017. https://ozone.unep.org/ozone-day/30th-anniversary-montreal-protocol-we-are-all-ozone-heroes.

United Nations Food and Agriculture Organization. 2010. *Food and Agriculture Statistics: FAO Statistical Yearbook 2010.* Accessed June 6, 2011. http://www.fao.org/economic/ess/ess-publications/ess-yearbook/ess-yearbook2010/yearbook2010-consumption/en/.

United Nations Food and Agriculture Organization. 2017. "'Energy-Smart' Agriculture Needed to Escape Fossil Fuel Trap." Media. Accessed September 3, 2017. http://www.fao.org/news/story/en/item/95161/icode/.

United Nations Human Rights Council. 2019. "Climate Change and Poverty." *Report of the Special Rapporteur on Extreme Poverty and Human Rights.* Accessed July 31, 2019. https://www.ohchr.org/EN/NewsEvents/Pages/DisplayNews.aspx?NewsID=24735&LangID=E.

United Nations Human Rights Office of the High Commissioner. 2019. "UN Expert Condemns Failure to Address Impact of Climate Change on Poverty." Accessed July 3, 2019. https://www.ohchr.org/EN/NewsEvents/Pages/DisplayNews.aspx?NewsID=24735&LangID=E.

United Nations Human Settlements Programme. 2003. "Slums of the World: The Face of Urban Poverty in the New Millennium?" Accessed June 18, 2012. http://www.unhabitat.org/pmss/listItemDetails.aspx?publicationID=1124.

United Nations Women. 1980. "World Conference of the United Nations Decade for Women: Equality, Development and Peace." Accessed November 20, 2013. http://www.un.org/womenwatch/daw/beijing/copenhagen.html.

United Nations Women. 2009. "United Nations Entity for the Gender Equality and the Empowerment of Women." *Convention of the Elimination of All Forms of Discrimination Against Women (CEDAW).* Accessed September 26, 2013. http://www.un.org/womenwatch/daw/cedaw/.

United Nations Women. 2011. "Fund for Gender Equality." Accessed June 18, 2012. http://www.unwomen.org/how-we-work/fund-for-gender-equality/.

United States Department of Agriculture. 2017. "The Role of Fossil Fuels in the U.S. Food System and the American Diet." Accessed September 3, 2017. www.ers.usda.gov/publications.

U.S. Department of Defense. 2014. "Climate Change Adaptation Roadmap." Accessed October 7, 2020. https://www.defense.gov/Newsroom/Releases/Release/Article/605221/.

U.S. Department of State Office of the Historian. 1997. *Foreign Relations of the United States, 1961–1963, Volume X, Cuba, January 1961–September 1962.* Accessed September 21, 2013. http://history.state.gov/historicaldocuments/frus1961-63v10/comp1.

U.S. Energy Information Administration. 2019. "How Much Oil Is Used to Make Plastic?" Frequently Asked Questions. Accessed July 11, 2019. https://www.eia.gov/tools/faqs/faq.php?id=34&t=6.

U.S. Immigration and Customs Enforcement. 2020. "Detainee Death Reporting." Immigration Enforcement. Accessed October 7, 2020. https://www.ice.gov/detainee-death-reporting.

U.S. Interagency Council on Homelessness. 2017. "Ending Chronic Homelessness in 2017: No One with a Disability Should Have to Experience Long-Term Homelessness." Accessed July 3, 2019. https://usich.gov/resources/uploads/asset_library/Ending_Chronic_Homelessness_in_2017.pdf.

Uzzell, David, Enric Pol, and David Badenas. 2016. "Place Identification, Social Cohesion, and Environmental Sustainability." *Environment and Behavior* 34 (1): 26–53.

Valerio, Nakita. 2019. "This Viral Facebook Post Urges People to Rethink Self-Care" *Flare.* Accessed July 8, 2019. https://www.flare.com/identity/self-care-new-zealand-muslim-attack/.

Vallina, Elvira Díaz, and Julio César González Pages. 2000. "The Self-Emancipation of Women." In *Cuban Transitions at the Millennium*, edited by Eloise Linger and John Walton Cotman, 15–32. Baltimore, MD: International Development Options.

Van Biema, David and Jeff Chu. 2006. "Does God Want You To Be Rich? A Growing Number of Protestant Evangelists Raise a Joyful Yes! But the Idea Is Poison to Other, More Mainstream Pastors." *Time Magazine*. Accessed October 8, 2017. http://content.time.com/time/magazine/article/0,9171,1533448,00.html.

Victor, David G. 2009. "Global Warming: Why the 2 °C Goal Is a Political Delusion." *Nature* 459 (18): 909.

Vigil, María López. 2008. "Heroines of the Special Period." In *A Contemporary Cuba Reader: Reinventing the Revolution*, edited by Philip Brenner, Marguerite Rose Jimenez, John. M. Kirk, and William. M. LeoGrande, 309–10. New York: Rowman and Littlefield.

Vitaliano, Peter P., Jianping. Zhang, and James M. Scanlan. 2003. "Is Caregiving Hazardous to One's Physical Health? A Meta-Analysis." *Psychological Bulletin* 29 (6): 946–72.

von Hassell, Malve. 2002. *The Struggle for Eden: Community Gardens in New York City.* Westport, CT: Bergin and Garvey.

Wackernagel, Mathis, Chad Monfreda, Dan Moran, Paul Wermer, Steve Goldfinger, Diana Deumling, and Michael Murray. 2005. "National Footprint and Biocapacity Accounts 2005: The Underlying Calculation Method." *Global Footprint Network: Advancing the Science of Sustainability*. Accessed October 5, 2013. http://www.footprintnetwork.org/download.php?id=5.

Wade, Lizzie. 2016. "Tesla's Electric Cars Aren't as Green as You Might Think." *Wired.* Accessed April 18, 2017. https://www.wired.com/2016/03/teslas-electric-cars-might-not-green-think/.

Walker, Liz. 2005. *Ecovillage at Ithaca: Pioneering a Sustainable Culture.* Gabriola Island, Canada: New Society.

Walk Free Foundation. 2018. "America's Report: The Global Slavery Index." Walk Free Foundation. Accessed November 14, 2018. https://www.globalslaveryindex.org/2018/findings/regional-analysis/americas/.

Wallace-Wells, David. 2019. *The Uninhabitable Earth: Life after Warming.* New York: Penguin Random House.

Wallerstein, Immanuel. 2004. *World-Systems Analysis: An Introduction.* Durham, NC: Duke University Press.

Waring, Marilyn. 1999. *Counting for Nothing: What Men Value and What Women Are Worth.* Buffalo, NY: University of Toronto Press.

Weaver, Richard M. 1948. *Ideas Have Consequences.* Chicago: University of Chicago Press.

Weber, Max. 1930. *The Protestant Ethic and the Spirit of Capitalism.* New York: Routledge.

Weinstein, Netta, Andrew Balmford, Cody R. DeHaan, Valerie Gladwell, Richard B. Bradbury, and Tatsuya Amano. 2015. "Seeing Community for the Trees: The Links among Contact with Natural Environments, Community Cohesion, and Crime." *BioScience* 65 (12): 1141–53. https://doi-org.proxy.lib.utk.edu:2050/10.1093/biosci/biv151.

Wessels, Tom. 2013. *The Myth of Progress: Toward a Sustainable Future.* Lebanon, NH: University Press of New England.

West, Stuart, Claire El Mouden, and Andy Gardner. 2011. "Sixteen Common Misconceptions about the Evolution of Cooperation in Humans." *Evolution and Human Behavior* 32 (4): 231–62.

Wichterich, Christa. 2015. "Contesting Green Growth: Connecting Care, Commons, and Enough." In *Practicing Feminist Political Ecologies: Moving Beyond the Green Economy*, edited by Wendy Harcourt and Ingrid Nelson, 67–100. New York: Zed Books.

Wiener, Cassandra. 2017. "Seeing What Is 'Invisible in Plain Sight': Policing Coercive Control." *The Howard Journal of Crime and Justice* 56 (4): 500–15.

Wiist, William H., Kathy Barker, Neil Arya, Jon Rohde, Martin Donohoe, Shelley White, Pauline Lubens, Geraldine Gorman, and Amy Hagopian. 2014. "The Role of Public Health in the Prevention of War: Rationale and Competencies." *American Journal of Public Health* 104 (6): e34–e47.

Williams, Raymond. 1973. *The Country and the City.* New York: Oxford University Press.

Wilson, Claire, and B. Moulton. 2010. *Loneliness among Older Adults: A National Survey of Adults 45+.* Knowledge Networks and Insight Policy Research. Washington, DC: AARP.

Winslow, Margrethe. 2005. "Is Democracy Good for the Environment?" *Journal of Environmental Planning and Management* 48 (5): 771–83.

Wood, Ellen Meiksins. 1972. *Mind and Politics: An Approach to the Meaning of Liberal and Socialist Individualism.* Los Angeles: University of California Press.

Woodward, Aylin. 2019. "The 'Lungs of the Planet' Are on Fire." World Economic Forum. Accessed October 7, 2020. https://www.weforum.org/agenda/2019/08/amazon-burning-unseen-rate/#:~:text=Plants%20and%20trees%20take%20in,oxygen%20in%20our%20planet's%20atmosphere.

World Bank. 2012a. *Turn Down the Heat: Why a 4°C Warmer World Must be Avoided.* Potsdam Institute for Climate Impact Research and Climate Analytics Report. Accessed April 12, 2017. http://documents.worldbank.org/curated/en/865571468149107611/pdf/NonAsciiFileName0.pdf.

World Bank. 2012b. *Worldwide Governance Indicators.* Accessed October 18, 2013. http://data.worldbank.org/data-catalog/worldwide-governance-indicators.

World Bank. 2017a. "GDP (current US$)." World Bank National Accounts Data, and OECD National Accounts Data Files. Accessed December 23, 2017. https://data.worldbank.org/indicator/NY.GDP.MKTP.CD?locations=CU.

World Bank. 2017b. "Household Final Consumption Expenditure, etc. (% of GDP)." World Bank National Accounts Data, and OECD National Accounts Data Files. Accessed April 14, 2017. https://data.oecd.org/hha/household-spending.htm.

World Bank. 2018. Groundswell: Preparing for Internal Climate Migration. Accessed October 7, 2020. https://www.worldbank.org/en/news/infographic/2018/03/19/groundswell---preparing-for-internal-climate-migration.

World Bank. 2021. "CO_2 emissions (metric tons per capita)." 1960–2016. https://data.worldbank.org/indicator/EN.ATM.CO2E.PC.

World Food Programme. 2017. "Cuba." Accessed August 25, 2017. http://www1.wfp.org/countries/cuba.

World Health Organization. 2017. "Depression and Other Common Mental Disorders: Global Health Estimates." Accessed August 25, 2017. http://apps.who.int/iris/bitstream/10665/254610/1/WHO-MSD-MER-2017.2-eng.pdf.

World Watch Institute. 2016. "The State of Consumption Today." Accessed June 28, 2016. http://www.worldwatch.org/node/810.

World Wildlife Fund. 2006. *Living Planet Report.* Washington, DC: World Wildlife Fund.

World Wildlife Fund. 2012. *Living Planet Report 2012.* Accessed October 5, 2013. http://awsassets.panda.org/downloads/lpr_2012_summary_booklet_final.pdf.

World Wildlife Fund. 2014. "Living Planet Report: Species and Spaces, People and Places." Accessed June 27, 2016. http://wwf.panda.org/about_our_earth/all_publications/living_planet_report/.

World Wildlife Fund. 2015. "Living Blue Planet Report 2015: Species, Habitats, and Human Wellbeing." Accessed June 27, 2016. http://www.worldwildlife.org/publications/living-blue-planet-report-2015.

World Wildlife Fund. 2016. *Living Planet Report 2016.* Accessed December 14, 2016. http://awsassets.panda.org/downloads/lpr_living_planet_report_2016.pdf.

World Wildlife Fund. 2020. "New WWF Report: 3 Billion Animals Impacted by Australia's Bushfire Crisis." Accessed October 7, 2020. https://www.wwf.org.au/news/news/2020/3-billion-animals-impacted-by-australia-bushfire-crisis#:~:text=Nearly%20three%20billion%20animals%20%E2%80%93%20mammals,birds%2C%20and%2051%20million%20frogs

Wright, Eric Olin. 2010. *Envisioning Real Utopias.* New York: Verso.

Yates, Michael D. 2012. "The Great Inequality." *Monthly Review* 63 (10). https://monthlyreview.org/2012/03/01/the-great-inequality/.

Yong, Ed. 2017. "How Coral Researchers Are Coping with the Death of Reefs: The Drumbeat of Devastating News Can Take Its Toll on the Mental Health of People Who Have Devoted Their Lives to Coral." *The Atlantic.* Accessed July 1, 2019. https://amp.theatlantic.com/amp/article/546440/.

York, Richard. 2006. "Ecological Paradoxes: William Stanley Jevons and the Paperless Office." *Human Ecology Review* 13 (2): 143–47.

York, Richard. 2007. "Structural Influences on Energy Production in South and East Asia, 1971–2002." *Sociological Forum* 22 (4): 532–54.

York, Richard. 2010. "The Paradox at the Heart of Modernity: The Carbon Efficiency of the Global Economy." *International Journal of Sociology* 40 (2): 6–22.

York, Richard. 2012. "Do Alternative Energy Sources Displace Fossil Fuels?" *Nature Climate Change* 2 (6): 441–43.

York, Richard, Christina Ergas, Eugene A. Rosa, and Thomas Dietz. 2011. "It's a Material World: Trends in Material Extraction in China, India, Indonesia, and Japan." *Nature and Culture* 6 (2): 103–22.

York, Richard, and Eugene A. Rosa. 2003. "Key Challenges to Ecological Modernization Theory: Institutional Efficacy, Case Study Evidence, Units of Analysis, and the Pace of Eco-Efficiency." *Organization & Environment* 16 (3): 273–88.

York, Richard, Eugene A. Rosa, and Thomas Dietz. 2003a. "Footprints on the Earth: The Environmental Consequences of Modernity." *American Sociological Review* 68 (2): 279–300.

York, Richard, Eugene A. Rosa, and Thomas Dietz. 2003b. "STIRPAT, IPAT and ImPACT: Analytic Tools for Unpacking the Driving Forces of Environmental Impacts." *Ecological Economics* 46 (3): 351–65.

York, Richard, Eugene A. Rosa, and Thomas Dietz. 2004. "The Ecological Footprint Intensity of National Economies." *Journal of Industrial Ecology* 8 (4): 139–54.

York, Richard, Eugene A. Rosa, and Thomas Dietz. 2009. "A Tale of Contrasting Trends: Three Measures of the Ecological Footprint in China, India, Japan and the United States, 1961-2003." *Journal of World-Systems Research* 15 (2): 134–46.

York, Richard, Eugene A. Rosa, and Thomas Dietz. 2010. "Ecological Modernization Theory: Theoretical and Empirical Challenges." In *The International Handbook of Environmental Sociology, Second Edition*, edited by M. Redclift and G. Woodgate, 77–90. Cheltenham, UK: Edward Elgar.

Young, Iris Marion. 1990. *Justice and the Politics of Difference.* Princeton, NJ: Princeton University Press.

Zalasiewicz, Jan, and Mark Williams. 2015. "Climate Change through Earth's History." In *Climate Change: Observed Impacts on Planet Earth*, 2nd ed., edited by Trevor Letcher, 3–17. Waltham, MA: Elsevier.

Zalasiewicz, Jan, Mark Williams, Alan Smith, Tiffany L. Barry, Angela L. Coe, Paul R. Bown, Patrick Brenchley, et al. 2008. "Are We Now Living in the Anthropocene?" *GSA Today* 18 (2): 4–8.

Zehner, Ozzie. 2012. *Green Illusions: The Dirty Secrets of Clean Energy and the Future of Environmentalism.* Lincoln: University of Nebraska Press.

Index

For the benefit of digital users, indexed terms that span two pages (e.g., 52–53) may, on occasion, appear on only one of those pages.